INDIAN COSMOGRAPH – A DISCOVERY

The fixed and the integrated from of the 27 stellar groups around the Earth

ROHINI SWAMINATHAN

INDIA · SINGAPORE · MALAYSIA

Notion Press

Old No. 38, New No. 6
McNichols Road, Chetpet
Chennai - 600 031

First Published by Notion Press 2018
Copyright © Rohini Swaminathan 2018
All Rights Reserved.

ISBN 978-1-68466-466-5

DEDICATED TO

My Parents, Family and Friends,
Rishi Bharadwaj and Ebenezer Burgess

CONTENTS

FOREWORD

Dr. R. Kannan, Ph. D., I. A. S.,
Agricultural Production Commissioner &
Secretary to Government, Agricultural Department,
Commissioner of Museums, Government Museum, Chennai-600 008
Phone: Off: 28193778/Email: govtmuse@md4. vsnl. net. in

After evolving the Cosmograph, the author showed the manuscript to his Head of the Department for his opinion, Dr. R. Kannan, Ph. D., I. A. S., Agricultural Production Commissioner & Secretary to Government, Agricultural Department, Commissioner of Museums, Government Museum, Chennai-600 008 wrote in his forward" Thiru. R. Swaminathan has not really wasted his time. Like the Rishis of old, he has speculated on the planets and stars, and discovered some ancient truths based on his knowledge of Surya Siddanta and traditional Tamil almanacs. I stumbled upon this, when during the course of my inspection, he suddenly showed me the manuscript for my opinion. This was because of my book on" Holistic Approach to dating in ancient History especially Indian History" published in 2000 AD by the Government Museum, Chennai. In that book, I have used astronomy and astrology to date events, which cannot be dated with certainty otherwise due to lack of material evidence.

Thiru. R. Swaminathan's theory on the positions of the Stellar constellations and the movements of the planets in the Solar system contradict modern astronomical theories. But that should not induce us to blindly reject it as incorrect. Dogma is the enemy of Truth. After all, only four centuries ago, the Europeans believed that the earth was flat. Galileo was killed for stating otherwise due to blind belief in currently held dogma. This was the work of religious fanatics but Europe had a Renaissance and the modern scientific temper and spirit of independent inquiry was born.

In the 19th and early 20th centuries, it was fashionable to criticize all ancient Indian beliefs. Now, even cow dung, the much derided object of that period, has been found many virtues and is used in agriculture as Panchagavya, a growth rejuvenator and biocide. Therefore, I will not be surprised if at some time in the future some aspects of the new theory propounded by Thiru. R. Swaminathan become standard reference." **I wish him and the book all success.**

Place: Chennai
Date: 10–01–2004

(Dr. R. KANNAN, Ph.D., I.A.S.)

FOREWORD

Prf. Dr. G. Devianayagam, 1-Arc (Italy) DSC., (h) Ph. D., (India) (Korea) Facu (UK).,
Head, Department of Architecture.
Tamil University,
Thanjavur.

India is the richest country in the world by possessing enormous resources of natural and cultural potentialities, the indigenous belief as well as the modern western scientific explorations is confirming that the first living being probably the Uni-Cellular Organism should have been evolved from this part of the Globe. Hence there is a long and continuous Heritage of lineage of knowledge System from the remote past penetrated through the time and still in living conditions is formed only in India. The society by Trial and Error Method attained the system of Knowledge and the very experiences reached by them were the determining factors of the polished, sophisticated wisdom of the Indians.

All these experiences otherwise called "The Wisdom" are deciphered to keep the Indian with a complete awareness of his life and of the environment around him. Hence the Indian wisdom is very well broadened in multiple disciplines by different involvement finally resulted as a capsule of "colloquy with Objects". The variations of the Objects were also very much understood by the sincere efforts of our ancient scientists called "Gnanis". These Gnanis employed themselves in search of the ultimatum of their own findings. By the way, search of the body and the soul of the mankind and the environmental materials were analysed. Therefore the like between the man and the universal factor like The sun, The Moon, Planets, stars of Akash, the Wind and the force, The fire, the water and the Land Mass with its Flora and Fauna were thoroughly investigated and understood. From which the sources to comfort the man were derived as fine methods of treatment of auvedh, Jyothis, etc. Money, Mantra and Aushath were also derived as fine methods of treatment of both the body and soul of the humanity.

Therefore it is easy to understand that the every step of knowledge was developed for the comfort to completeness of life of the human society. Moreover it gives the clear idea that almost all the existing objects were deeply investigated and inferred by our Indian Gnanis beyond instrumentation (other than Human body). The Vijanm of the Indian tradition were classified under different fields and The Joyothisha were one among them. Joyothisha is the scientific study of the space around the earth by which the celestial objects were thoroughly deciphered and understood. When this study was used in the applied science on the life of the man it was differently understood from the physical study of the celestial beings. Hence the astronomy was identified as Asrtrology only by astronomical applications to study the life of Mankind. In making the study on the celestial lighting object upon the earth Indian Gnanis realized the determining fundamental factors of observation and understanding, they are nothing but the everlasting resources of space and time. The variance of these spaces gave birth to different schools of "Akash Sastra". One of our celebrated scholars of our ancient astronomy was Aryabatta. The other one is "Varaha Ora".

The observations and understandings of these Gnanis were in practise among the Indians only by the oral tradition in the early period and at the later point of time were recorded in palm leaf in manuscript Forms. Number of books from by different authors from different perspectives evolved. Every one of them was genuine and complete in their own accord. One such book is Surya Siddantha.

The entry of the Europeans in India by the Christian Missionaries in the 16th century bought many enquiries over the crystal clear application of an indigenous knowledge system, As sceptical they were making some efforts to understand the local tradition after a good experience resulted to go for the translation of the Indian Texts. Such an attempt of translation by Ebenezer Burgess falls on Surya Siddhantha which is the source book of Mr. SWAMINATHAN of Thanjavur to make this research book-Let.

Mr. SWAMINATHAN is an experienced person in the field of ancient building art and very well acquainted with the jyothisha and environmental studies. He is industrious Scholar of Traditional sciences of Indians and having a strong belief of them. His studies over the Indian approach of the celestial objects were beyond any enquires. Therefore with his understanding of indigenous knowledge of Astronomical studies especially on the studies of the planets, he wants to present the identical facts of reality, realized and understood by our ancestors to the present generation by the present article. The Indian Cosmograph-A

Discovery-this booklet is nothing but an attempt of explorative study on the Indian expertise over the Astronomy.

In this booklet, he tried to present the pictorial representation of the position of 27 stars and 9 planets in the Cosmos with the traditional knowledge. Accordingly the stars and the planets are in arched lines above the earth in different layers. The Earth is the centre of the Universal Equator, The Indian concept of Raghu and Kethu are nothing but the powerful thick shadows of the Planets, Saturn and Mars respectively. The Kuligan should be the shadow of the Jupiter. There are three influencing shadow Planets as the other ones are negligible and the possibility of shadow is a very meagre.

The western concept of The Sun and Moon eclipses are invalid whereas the Indian contexts of these eclipses were praise worthy. The Sun and the moon eclipses happen only because of the shadow planets called Raghu and Kethu respectively by their movements. This has been understood by a critical point of 180 degree of a line of Imagination where all the sun, moon, Ragu or Kethu and the earth will be positioned. Therefore the understanding of the cosmic energy is also totally different from the western which is an important study and very much needed for the present world.

By this way of exploration Mr. SWAMINATHAN is not simply making his presentation by words but also presenting by sketch, drawings which deserves for appreciation, The author gives the vivid account of Indian concept of Indian Cosmology so lucidly. It is my fervent faith at the outset there will by only blunt rejection, but a thorough study of this paper will definitely make the reader to go beyond the instrumentation where he can understand the eternal facts of the cosmos.

06.05.2003

Thanjavur

EDITORS NOTE

I am delighted to write the foreword for the research work Indian Cosmograph-A discovery and really honoured to be part of the editorial team of this wonderful attempt. Mr Swaminathan in the early 2000s after admiring the Surya Siddanta came forward with this wonderful thought of exploring the Ancient Indians astronomical treatise and contributed to the Indian society in briefing the values of the ancient Indian Astronomical theories. This research work provides vital information about astrology followed by Ancient Indians, planets and stars—aspects of our environment which are unseen but strongly felt. With this knowledge at our disposal, we come to understand the characteristics of the planets and the stars and their influence to the day to day activities of all beings in the earth and also outside the earth through the entire cosmos

The chapters in this research work include Astrology, and Astronomy of western concepts and that of the Ancient Indians based on which the detailed study of how they arrived the order of the weekdays, months and Year and the entire calendar system. Each chapter contains evidence-based background information from almanac (Panchangam), which are intended for the professional who already possesses a basic understanding of the Astrology and Ancient Indian Astronomy. The concluding part have set of annexures with detailed information for the reader's to have better understanding of the subject, It is my hope and expectation that this book will provide a detailed and excellent learning experience and referenced resource for future generation in the subject of Ancient Indian Astronomy.

Mr. Swaminathan's research work on the positions of the 27 Nakshatra groups and the movements of the planets in the Solar system will be discrepant to the theory that we follow in the current astronomical practices. But after having a thorough understanding of this research work and the detailed explanation given by him in the form of Graph one should not deny the work. I personally appreciate the author's effort in bringing the astronomical treatise followed by

Ancient Indians to the notices of our Government. His approach to the various Governmental bodies to include the same in the educational system and his Public interest Litigation filed before the Honorable justice in the High Court of Madras, has paid him a result and court ordered the concerned department to look in to the subject and consider the same to include in the educational system. I wish him all success towards the work and request him to continue his research in bringing in the values and practices followed by our Ancients scientifically.

Hariesh Kumar N, BE EEE, MBA IB (Singapore),
Senior Manager (Banking Industry)
Chennai.

PREFACE

In Ancient India, the 64 main branches of Science such as Agriculture, Architecture, crafts, Dance, Drama, Education, Geology, History, Iconography, Jyothisha, Rituals, Music, Medicine, Politics, Religion, Yoga and so on were developed on the basis of "Oneness of the Earth centred Cosmic theories". The Ancient Indians derived their Oneness of the Earth centred cosmic theory from the fixed and the integrated form of 27 Nakshatra groups (stellar) around the earth in ascending and descending order. This forms the shape of the Cosmos which is "Hollow egg" in shape, otherwise called as "ANDAM". Within the hallow space of the Cosmos, between the Nakshatra (stellar) Groups and the Earth, the Planets are positioned eccentrically at different locations. The Earth is positioned at the centre of the cosmos and the movement of the planets refer to the fixed and the integrated form of 27 Nakshatra groups (stellar) which are seen revolving around the Earth. The movement of the planets and the stellar energy makes the Earth to be fixed at the centre of the cosmos without any movement unlike other planets in the cosmos.

Ptolomy, an Egyptian Astronomer, Sailor, has the same opinion as Ancient Indians, that the Earth is at the Centre of the Cosmos. The Ancient Indians and others who believes that the Earth is at the Centre of the Cosmos had failed to their claims to the then Astronomers, Aristotile, Galileo, Copernicus, Kant and Keplar who all derived a model of the universe that the Sun is at the centre of the Cosmos rather than the Earth. They also arrived to a model that the Planets including the Earth are revolving around the Sun in elliptical Orbits. Thus by changing the order of the Ancient Indian Planetary positions, and by assuming the "Planet Moon" as the satellite Planet to the Earth without conferring to the position of the stellar groups in the Cosmos.

Ebenezer Burgess, in his translation of Surya Siddhantha, an ancient Indian Astronomical Treatise, chapter viii P, 209 "Of the Asterisms" Writes, "The number and configuration of the Stars forming the groups is not stated in our text; we derived them mainly from Colebrooke, although ourselves also having had access to and compared, most of his authorities, namely Cakalya-Sanhita, the Muhurtha

Cinthamani and the Ratnamala (as cited by Jones As. Res. ii 294) Sir William Jones, it may be remarked, furnishes (As. Res. ii, 293 plate) an engraved copy of drawings made by a native artist of the figures assigned to the asterisms. For the number of Stars in each group we have an additional authority in al-Biruni, the Arab servant of the eleventh century, who travelled in India, and studied with especial care the Hindu Astronomy. Al-Biruni also gives an identification of the asterisms, so far as the Hindu Astronomers of his days were able to furnish it to him, which was only in part, he is obliged to mark seven or eight of the series as unknown or doubtful. He speaks very slightly of the practical acquaintance with the heavens possessed by the Hindus of his time, and they certainly have not since improved in this respect; the modern investigators of the same subject, as Jones and Colebrooke, also complain of the impossibility of obtaining from the native Astronomers of India satisfactory identification of the asterisms and junction of stars. The translator, in like manner, spent much time and effort in the attempt to derive such information from his native assistant, but was able to arrive at no results which could constitute any valuable addition to those of Colebrooke. It is evident that for centuries past, as at present, the native tradition has been of no decisive authority as regards to the position and composition of the groups of stars constituting the asterisms; these must be determined upon the evidence of more ancient data handed down in the Astronomical Treatises".

Also Burgess, in his translation of Surya Siddhantha, chapter viii P, 242 "Of the Asterisms", "Much yet remains to be done, before the History and use of the system of asterisms, as a part of the ancient Hindu Astronomy and Astrology shall be fully understood. There is in existence an abundant literature, ancient and modern, upon the subject, which will doubtless at some time provoke laborious investigation, and repay it with interesting results. We have already allotted to the Nakshtras more space than to some may seen advisable our excuse must be the interest of the History of the system, as part of the ancient history of the rise and spread of Astronomical Science; the importance attaching the researches of M. Biot, the inadequate attention hitherto paid them, and the recent renewal of their discussion in the Journals des savants, and finally and especially, the fact that in and with the asterisms is bound up the whole history of Hindu Astronomy, prior to its transformation under the overpowering influence of western science".

Though the Indians are practising their Ancestors' Oneness Universe, The Earth Centred Cosmic Theory in their day to day affairs using the Almonacs (PANCHANGAMS) which are briefing the detailed movement of the planets along and parallel to the fixed and the Integrated Form of the 27 Nakshatra (stellar) groups around the earth in the cosmos. But no Indian Astronomer or

Astrologer of India had provided the location of the Stellar groups around the Earth with reference to their Panchangam, or any other Astronomical treatises. Now, after having a detailed and careful study of the Indian Astronomical treatises and the Panchangam, the location of the fixed and the Integrated Form of the 27 Nakshatra (stellar) groups around the earth in the cosmos have been deciphered. The Astronomical Events such as formations of eclipses, the phases of the Moon and the 8 kind of movements of the Planets with reference to the fixed stellar groups in the Cosmos have been explained and is entitled as **"Indian Cosmograph – A discovery"**. This Cosmpgraph can be updated to any time and can visualise the exact location of the planets around the earth in the Cosmos.

Before going in to the deciphered Model I Sincerely acknowledge and thankful to the under mentioned for their concern on the research and for their valuable suggestions, Thanks for the guidelines and the prefaces to, Dr. R. Kannan, Agricultural Production Commissioner & Secretary to Government, Agriculture Department, Commissioner of Museums, Government Museum, Chennai, Prof. Dr. G. Deivanayagam, Head, Department of Architecture, Tamil University, Thanjavur., Sri. T. S. Jaganathan, Southern Railway, Srirengam, Sri. Srinivsan (Pachaikkal), Retd., Supdt., Kerala State P. W. D, Mannargudi, Shri. S. Paranan, Curator, Department of Archaeology, Thanjavur, Sri. R. Arunajadesan, UCO Bank Retd., Chennai, R. Vaidyanathan, Retd., Tamil Pandit, The Madras Progressive Union Higher Secondary School, Chennai, Smt. Komalavalli, W/o Late Narasiman, Mannargudi, Mrs. Rohini swaminathan, PET., Government Higher Secondary School for the vision impaired, Thanjavur, R. Umamaheswaran, Department of Archaeology, Thanjavur,

My sincere thanks to Advocate Raghunathan, Chennai to come forward to file a public interest litigation writ petition seeking the directions of the Chennai High Court to include the Indian Panchangam and the abundant Indian Astronomical treatises in the Educational Institutions in India and obtained successful orders from the High-Court, Chennai.

INTRODUCTION

"Sarvesham upari
nakshthrani thadhadou sanaicharaha
thadhathou Guruhu
thadhadhav Bowmaha
Thadhatsh Suryaha
Thathadus sukraha
Thadhathou Budhaha
Thathadhas chandraha Ethigrahanamkakshakramanaha"

The present concept of the Solar System is not apparent to what the Indians practised earlier, the basic concepts of the cosmic principles found in ancient Indian Astrology are practiced as ceremonies in our day to day life. It is understandable that the vedic principles of ancient Indians would have been derived after finding the fixed positions of the stars and the movement of the planets under their influences around the earth in the Universe. Moreover the picturization of the Universe is well defined in "Surya Siddhatha" in the form of verses, vakyas and slokas

The orders of the weekdays are derived from the names of the planets which are positioned in the stellar groups that cover the cosmos. The "sukala yajur Vakya" mentioned above describes the order of the Planets in the hallow-space of the cosmos between the stellar-groups and the Earth. To derive the detailed Calendar systems, Ancient Indians differentiated the routine and consecutive sunrise over the Earth, for that purpose they utilized the order of the Planets in the cosmos, and by using the theory of Hora, they arrived the order of week days as 1. Saturday 2. Sunday 3. Monday, 4. (Mars) Tuesday, 5. (Mercury) Wednesday, 6. (Jupiter) Thursday, and 7. (Venus) Friday.

Then they calculated Swasa to Vinadi, Nazhigais, Day and Night, 15 days, or 2 pakshas, 12 months, 2 Ayanaas (Uthraayana and Dakshina ayana) which is 60 cycle of years and 1 Deva day, 1 Deva year, 4 Yugas, 14 Manvantras, and 8 Kalpas. The 12 months or the 2 Ayanaas are classified from the movement of the Sun from its position along and parallel to the Nakshtra's energy-line created in the Cosmos. The Sun transit parallel to the fixed stellar-groups in the Northern Cosmic hemisphere is known as Uthra-Ayanaa, and the Sun transit in Southern Cosmic hemisphere is known as Dakshina-Ayanaa.

The system of Indians which differentiate the sequence of sunrise over the Earth is known as HORA. In India even to-day the HORA system is in practise. Prof. H. H. Wilson expresses, "The origin of the arrangement is not precisely ascertained, as it was unknown to the Greeks, and not adopted by the Roman until a late period. It is commonly ascribed to the Egyptians and Babylonians, but no very sufficient authority, and the Hindus appear to have at least as good as little to the invention of any other people."

The ancient Astronomers of China, India and Arab have identical views, while Babilonions and Greek have different perceptions and the Westerners differing from both, the names of the Months is also derived by the Ancient Indians using the stellar-group names, when the planet Moon transit through a particular Stellar-group in Full-Moon status, For example, when the moon transit through the stellar group (14) Chitra which is taken as the name of the first month in the Indian calendar system.

Ebenezer Burgess States, "In the Modern astronomy of India, the Nakshatras (Stars) are of subordinate consequence only and appear as hard more than reminiscences of a former order of things: From the Surya Siddhantha might be struck out every line referring to them without serious alteration of character of the treatise." Also he pointed out from Jour, roy, (AS, SOC, ix, 84) that Hindus appears to have at least as good a title to the invention as any other people. "One word on the claims of the Arabians to the honor of original invention is Astronomical Science. And first they themselves claim no such honour. They confess to having received their Astronomy from India and Greece. They were thoroughly imbured with the knowledge of Hindu Astronomy before they became acquainted with that of the Greece, is evident from their translation of Ptolemy's Syntaxis. During the Eleventh Century, Al-Bruni also gives an identification of asterisms so far as the Hindu Astronomers of his days were able to furnish it to him which was only in part, he is obliged to mark 7 or 8 of the series as unknown and doubtful. The modern investigators Sir William Jones and Colbrooke also complain of the impossibilities of obtaining from the native Astronomers of India satisfactorily in identification of the asterisms and their junction of Stars".

From The above statements, It is evident that the Indians are the Pioneer of Astronomy which inspired me and took me through the Indian Panchangam. Astronomical treatise of India and Indian Astrological theories which deciphered the fixed and integrated form of 27 Stellars, Asterisms and the junction of stars around the earth in the Cosmos and further the work is entitled as The Indian Cosmograph – A Discovery.

CONCEPT OF UNIVERSE

1. 1 WESTERN CONCEPT

Philosopher Kant (1755 A. D.) expounded the theory of solar system which was systematized in 1796 by Laplace on mathematical grounds. This system asserted the Sun is stationery and around which all planets revolve in orbits in the same plane, their movements being east to west. Keplar derived the laws regarding the motion of the planets. And he also explains the phenomenon of eclipses and New moon. May be these were the foundations to enunciate the coincidence of all the orbits of the planets and the equator in the same plane. There is a conspicuous absence of any reference regarding the position of stars in layers and movement of the planets in accord with relation to the stars positions as shown in the fig.

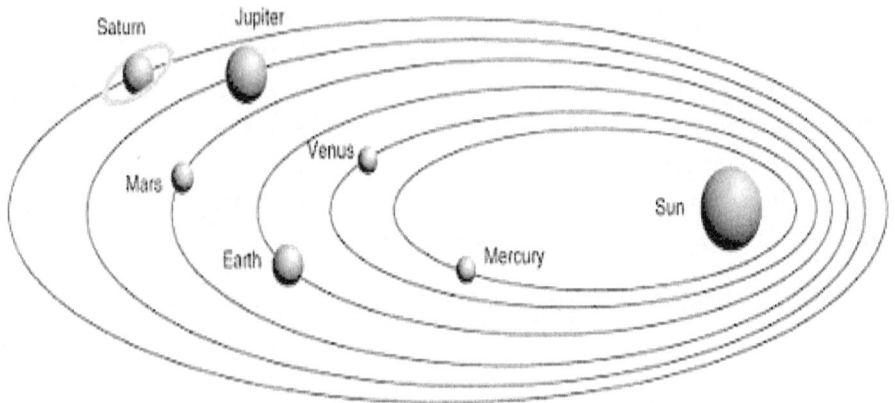

Fig: Kepler's Law of Universe

1. 2 INDIAN CONCEPT AND THE INDIAN PANCHANG

The Geo-centrists in the past all over the globe believe that the Earth is center of the Cosmos which has no proper refernce even after several investigations and studies.

Ebenezer Burgess, a British Civil Administrator to India in Mumbai, in his Surya Siddhantha wrote," Much yet remains to be done, before the History and use of the system of asterisms, as a part of the ancient Hindu Astronomy and Astrology shall be fully understood. There is in existence an abundant literature, ancient and modern, upon the subject, which will doubtless at sometimes provoke laborious investigation, and repay it with interesting results".

Eventually, a labourious and a detailed study of the Nakshatras with the Indian Calendar/Panchang and the Astronomical Treatise, the Surya Siddantha, reveals that the Earth is positioned at centre of the Cosmos and all the 27 Nakshtra groups surrounds the Earth in Ascending and Descending order with reference to its Equator and its pole line. An Indian Vakiya explains the order of the 7 Planets which are positioned in the space between Nakshtra groups and the Earth in the cosmos. The orders of the seven Planets from the Nakshtra groups are as follows, 1. Saturn, 2. Jupiter, 3. Mars, 4. Sun, 5. Venues, 6. Mercury, and 7. Moon.

To arrive the detailed Calendar systems, initially ancient Indian were in need of differentiating the routine and consecutive sunrises over the Earth, for this purpose they utilized the order of the Planets in the cosmos, and the theory of Hora, basis which they arrived to the order of weekdays as 1. Saturday, 2. Sunday, 3. Monday, 4. (Mars) Tuesday, 5. (Mercury) Wednesday, 6. (Jupiter) Thursday, and 7. (Venus) Friday.

Then they classified the beginning time which starts from 1 Swasa to Vinadi, Nazhigais, Day and Night, 15 days or 2 pakshas, 12 months, 2 Ayanaas (Uthraayana and Dakshina ayana), 60 cycle of years, 1 Deva day, 1 Deva year, 4 Yugas, 14 Manvantras, and 8 Kalpas. The 12 months and the 2 Ayanaas are classified on the basis of the movement of the Sun from its position along and parallel to the Nakshtra's energy-line created in the Cosmos.

1 nazhigai = 6 degree rotation of the earth

24 hours = 60 nazhigai

1 hour = 2. 5 nazhigai

1 nazhigai = 24 minutes

Thus, the Sun transit parallel to the fixed stellar-groups in the Northern Cosmic hemisphere is known as Uthra-Ayanaa, and when the Sun transit in Southern Cosmic hemisphere is known as Dakshina-Ayanaa.

The 60 cyclic years of Indian Panchang/Calendar, clearly shows the detailed movements of all the 7 Planets and inclusive of the Shadow Planets with reference to fixed and the integrated form of the 27 Nakshtra groups around the Earth in the Cosmos with 8 kinds of movements from time-to-time. Further, the Panchangam shows accurately the time of occurrence of the Sun and Moon Eclipses also it details which shadow planet Rahu or kethu which is the cause of the eclipse in Sun or Moon and the color of the Eclipsed planet.

INDIAN-COSMOLOGY

The Indian Panchangam provides the very important and the very basic data about how the 27 Nakshtra groups are spread over the Earth in the Cosmos. Since the Earth is positioned at the centre, the extended Earth's equator, which acts as the Cosmic-equator and the extended Pole-line of the Earth as the Cosmic-pole-line. The Panchangam details the starting points of the stellar-groups from the cosmic equator. The first-stellar groups begin at 0deg. and ends at 13deg. 20min at the first-layer in ascending order. The 2nd stellar-group starts at the 13deg. 20min. in ascending order and reaches the 2nd layer and ends at 26deg. 40min. The 3rd Nakshtra-group begins at 26deg. 41min. in ascending and reaches at 40 deg. at 3rd layer.

The 4th group in ascending order and reaches at 53deg. 20min, the 5th group horizontally with no ascending or descending and ends at 66deg. 40min. The North Cosmic Pole ends at this level. The 6th group starts here and reaches 80deg. in descending order. The 7th descends and reaches 93deg. 20min. and act as the connecting the Northern hemi-Sphere to the Southern hemi-sphere of the Cosmos. The 8th group starts here in descending order and reaches at 106deg. 40min. The 9th group begins here and ends at 120deg. At 120deg. the 10 the group starts in ascending order in the Southern hemi-sphere and the 11th, 12th, 13th group of nakshtras reaches at 173deg. 20min. The 14th group 173deg. 20min-186deg. 40min spreads horizontally at the Southern most edge of the Cosmos. From this point the 15th, 16th, 17th and the 18th groups in descending order and reach at 240 deg. From this point the 19th, 20th, 21st and the 22nd groups followed consequently and reaches in ascending order at 293 deg. 20min, where the 21st groups spreads parallel to the Cosmic pole. The 23rd group spreads between 293deg. 20min.-306deg. 40min with no ascending or descending but parallel to the Cosmic Equator. The 24th, 25th, 26th and the 27th stellar groups descend from 306geg. 40min and reach the 0deg, where the 1st group start in the Cosmos. The Surya-siddhantha, an astronomical text shows that the ascending and descending ordered 27 Nakshtra groups covered Cosmos looks like an hallow Egg or 2 bowls placed one over the other with mouth thus creating a hallow space within.

In the hallow-space created by the fixed and the integrated form of 27 Nakshtra groups i. e. in the Egg-shaped Cosmos, the Earth is positioned at the centre of the Cosmos due to the Nakshtra's energy and self-rotating movement. Due to the excess energy from the Nakshatra group the centre between the 7th and the 21st groups which are spread over and parallel to the Cosmic Pole. The other 7 planets, due to their eccentric placements have their 8 kinds of movements along and parallel to Nakshtra's positions in the Cosmos. The Surya-siddhantha reveals that the circumferential measurements of the integrated form of the nakshtra's and the 7 Planets in the Cosmos, are in Yojanas unit which are to be reviewed with the modern linear-units. But no circumferential measurement for the Earth is provided, hence it is at the centre of the Nakshtra groups which covers the Cosmos.

Further, the ancient Indians with their understanding that the Planets below the Nakshtra's are the Planet Saturn, Jupiter and the Mars which have casted their shadow over the Earth, due to the stellar energy in the Cosmos. Thus the shadows formed from these planets, have their movements in the anti-direction of their parent planets that is in anti-clock-wise direction in the Cosmos, which causes the Solar and Lunar Eclipses in the Cosmos. These are fully explained in the Indian-Panchangam from time to time for 60 cyclic Years.

From the above understandings of the Ancient Indians, they derived their cosmological-theories over the Oneness Universe created with the fixed and the integrated form of the 27 Nakshtra groups that surrounds the Earth in the Cosmos. They clubbed the 27 Nakshtra groups into 12 Constellations. Taking the movement of the Sun with reference to the Nakshtra constellations they named their 12 months. When the Sun transits in its position, parallel to the constellation groups in the Northern Cosmic hemisphere it is designated as Uthra-Ayana and when it transits in Cosmic Southern hemisphere is known as Dakshina-Ayana.

It is learnt from the detailed study of the Indian-Panchang and the Surya-siddhantha that the Ancient Indians Oneness Universe theories are derived on the acceptable scientific facts. The Ancient Indians Nakshtra based Oneness Cosmic Theories will not get immediate acceptance by going through a glance and will be appreciated only after a deep study and discussions beyond any religious faiths which will be instrumental to fulfil the desire of Sri. Ebenezer Burgess.

ASTRONOMY

With the advent of civilization the thinkers started venturing into the realm of Nature to understand the natural phenomena such as day, night and rain etc., The great thinkers of the ancient time with their close and keen observations of things around have, had come to certain conclusions. One of the branches of their study centred around the space, heavenly bodies such as stars and planets and their relationship to earth and life on the planets, The Indians, Chinese, Arabs, Babylonians and other western thinkers had formulated their findings and thus various treatises are available on this subject of Astronomy. However, experts in this field like Cole Brook and Ebenezar Burgess are of the view that Indians were pioneers in the field of Astronomy and foremost on expounding the subject.

Burgess feels that the position of constellations could not be precisely understood in the Hindu System and further studies of the Hindu astronomical concepts could reveal more information and lead to more enlightenment. Though he feels that the Hindu system of astronomy has relegated the positioning of stellar constellations to a lesser important place, the fact remains that the Hindu life style, which is now equated with Hindu religion is based on the fundamentals of the astronomical concepts of the ancient Hindus. This review is a humble attempt to understand the rationale of the basic doctrines of the ancient Hindu Astronomers vis-a-vis the Western concept of solar system on the basic of the translation by Ebenezar Burgess of the Indian treatise, the "Surya Siddhantha". My recreated diagram based on the Hindu Cosmology attached herewith is also based on the measurements furnished therein.

COSMOLOGICAL INVENTIONS BY ANCIENT INDIANS

INDIAN sub-continent on the Earth is identified by the other part of the World with her Mount Everest in the Himalayas with the Perennial, spiritual river Ganga and with her Spiritualism. She has the richest Heritage and well known for her Yogic and spiritualistic way of life. Her children bestowed with WISDOM OF KNOWLEDGE, who have contributed many inventions and discoveries in all lifestyles! Among them, the cosmological inventions and discoveries made by them are the standing examples for their excellence of knowledge, which played a key-role to lead a peaceful life with humanity!

In the modern times, the spiritualistic approach of life considered, as an unscientific who has only unfolded believes of bad faiths. Such believes, which cannot be reasoned easily are considered as the spiritual or bad-faiths. If we can through of such bad faiths, we can find out the proper reasoning based upon the Ancient Indians theories of the Cosmological Inventions and Discoveries.

ANCIENT INDIAN COSMOGRPAHY

"Sarvesham upari nakshthrani thadhadou
Sanaicharaha thadhathou
Guruhu thadhadhav Bowmaha

Thadhatsh Suryaha Thathadus sukraha

Thadhathou Budhaha Thathadhas chandraha

Ethigrahanamkakshakramanaha"

The above sujla yajur samhitha vakyam describes the positions of the stars and planets in the Universe. In the universe the first layer is the star groups and then the planets Saturn, Jupiter, mars the Sun, Venus, Mercury and Moon. As the Vedic studies of this branch were taught orally from generation to generation and practiced accordingly, the date of first documentation is not clear. Tradition says, that 18 Siddhanthas or doctrines on the subject were in circulation. The "Surya Siddhantha" is claimed to be the oldest and earliest of all the doctrines. As per the above Siddhantha, we get a pictorial view of the universe as described below.

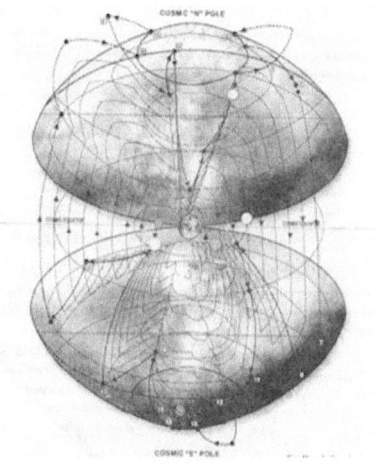

Fig: Pictorial View of the Universe

The Universe looks like an egg or two bowls placed one over the other with wide open at the centre, thus creating a hollow space within. This hollow space is filled with Maha-Prana, the energy transmitting media where the top most region is filled with the stars in multiple layers. Then the seven planets follow the same in the order of Saturn, Jupiter, Mars, Sun, Venus, Mecury, Moon and the Earth in the descending order (Earth is not taken in to account in the above pictorial representation). The stellar constellations are in a static position at different levels and the planets move in their own orbit with the earth at the centre point in a stationary position revolving on its own axis. The planets are moving in North-South and South-North directions having a total displacement of East to West within the extremes of their respective orbits of longitudes 0° to 360°.

Slokas 29 to 90 section 12 of "Surya Siddhantha" deals with the above factors. While furnishing the order of the planet it details the circumference of the universe, the orbits of the planets, as indicated below.

Universe	18, 112, 080, 864, 000, 000	Yojanas
Stars	259, 890, 000	Yojanas
Saturn	127, 688, 285	Yojanas
Jupiter	51, 375, 764	Yojanas
Mars	8, 146, 909	Yojanas
Sun	4, 351, 000	Yojanas
Venus	2, 664, 637	Yojanas
Mercury	1, 043, 209	Yojanas
Moon	324, 000	Yojanas
(1 Yojana = 5 miles or 8. 045 Km)		

5. 1 THE SOUL

The Ancient Indians deciphered the difference between the *MORTAL AND IMMORTAL*. They specified the mortal is the constituent of Worldly thing, which formed from time-to-time due to the chemical and physical changes, and the immortal is the part of the cosmic energy, which is generated by the 27 Nakshatra Groups and the planets in the Cosmos.

They designated the Soul as the part of the cosmic energy which rotates on the earth with reference to the energy prevailing in the cosmos from time-to-time based on the position of the planets with reference to the earth in the Cosmos. Therefore, the Ancient Indians defined the soul as immortal. The Soul which has a temporary Shelter when it is connected to the mortal body made of the Worldly things with the impact of the cosmic-energy from-time-to-time for their growth and destruction. With their Geo-centric Oneness Universe made out of the fixed and the integrated form of the 27 Stellar, they deciphered that the immortal Soul have numerous contacts from time-to-time with the mortal bodies in a cyclic pattern due to the influence of the Cosmic Energy and had claimed the theory of Re-birth. Moreover they found out the relation between the worldly things with the cosmic energy with reference to their Physical and Chemical qualities in the Sphere of Earth.

The Cosmic-energy, which plays a vital role on all the substances of the earth which attracts and distracts and it is the major cause for their similarities and differences. On visualizing the above, they found-out that the Souls are immortal in nature and they are the part Cosmic energy, whereas all the other perishable maters over the earth are formed with the combinations of the Five Factors of energies which are Air, Water, Heat, Earth and the Cosmos.

COSMOLOGICAL DISCOVERIES
AND INVENTIONS IN INDIA

Ancient Indians were the pilots over the Earth to unfold the Secrets of the Cosmos, which made out of the Fixed and the integrated form of the 27 Nakshatra groups through their perishable bodies as a Model of the Cosmos. They undertook a detailed study about the cosmic energy and its influence over the perishable body and it's Soul. The *Rishies and the Siddhars of India* to extend their life-time practiced *Yoga* and repeated a particular sound that activates the cosmic energy around them than a normal man. Some of them tried through the Herbals and thereby developed the Ayurvedic and Siddha System of Medicine in India.

Also Ancient Indians identified that the systematic pronunciation of a sound which can vibrate the Cosmic energy based on which they developed the system of Mantras. Further, they deepened their study and identified that the different type of sounds have different types of frequencies through which anything can be identified easily on the Earth with reference to the Cosmic Energy. They understood about the direct and the indirect influences of the cosmic energy over the each matter on the Earth. Further, they found that the Lights, which are emitting from the stellar or the Nakshatras and from the planets have also their influence over the Matters of the Earth which are seen in different colors.

Also ancient Indians after having a thorough study on the environmental behaviours around them noted a systematic and cyclic rotation over a span of time which is mainly due to the movement of the Earth and the movements of the Planets with reference to the Fixed and the integrated form of the 27 Stellar or Nakshatras in the Cosmos. Finally, they come to conclusion on the exact locations of the 27 different Nakshatras groups with reference to the Earths' Equator and its pole-line compared to the structure of the Human body. This identification of exact locations of the 27 different Nakshatras groups has an integrated form from Foot to Head Using the Human structure as the model of the Cosmos. They deciphered that the 27 different types of Nakshtra groups are spread over in the cosmos with reference to the Earths' Equator in Ascending (Arohana), and in

Descending (Avarohana) order and complete their integrated form at the Starting point at 0″ position. These Integrated form of the 27 Nakshatra groups which are positioned at the outer most layer of the Cosmos and the energy generated by them are filled within the Cosmic-Sphere surrounded by them.

The Cosmic-Sphere found is filled with the cosmic energy and the energy from the nakshatra group causes the fixed position of the Earth at the centre of the cosmos. The excess residual energy derived at the centre of the cosmos by the Nakshatra groups causes the Earth to have a constant spinning movement at the centre of the cosmos. The other planets have their movements along and parallel to the Fixed and the integrated form of the Nakshatras as they are positioned in eccentric positions in the Cosmic Sphere.

6. 1 THE FIXED POSITIONS AND THE INTEGRATED FORM OF THE 27 NAKSHATRA GROUPS IN THE COSMOS

The Ancient Indians, initially derived their exact living locations on the sphere of the Earth by sub-dividing the same with Atchamsa rega and Deergamsa rega (longitudes and latitudes) using the Bhumadhiya rega and the Duruva rega (Equator and the Pole-line) as the base lines to locate their exact location. To locate the exact positions of the Planets and the Positions of the 27 nakshatra groups in the Cosmos they extended the Latitudes, Longitudes, the Equator and the Pole-line of the Earth and named it as Cosmic Equator and the Cosmic Pole. Then they bisect the Cosmos into Northern and Southern Hemisphere based upon the Cosmic Equator. To fix the Starting and the ending positions of the each Nakshatra groups at different levels from the Cosmic-equator, along the pole line in ascending and descending order, they referred them to the structure of the normal human-body, which has four main levels of foot, thigh, stomach, neck and head.

They invented that the First Group of Nakshatra which has the starting point at the foot level which is found to be parallel to the Cosmic equator at 0 deg. and spread over in the Cosmic Northern sector and ends at 13°20″ in the ascending order and ends at the second layer where the thigh portion begins. Following this, the next second group of Nakshatra starts at 13°20″ and finishes at the 3rd layer at 26°40″ at the Stomach level in ascending order. On continuing the 3rd, the 4th Nakshatra group starts at 26°40″ and ends at the 4th layer, at the neck at 40° and ends at the head at 53°20″. The 5th Nakshatra group which spread at the Northern Cosmic sphere which has no Ascending or Descending order but in a part of circular arch portion between 53°20″and 69°40″ parallel to the Cosmic Equator.

16

From the outer most Northern Hemisphere where the 5th Stellar group spread over in row Cosmos keeping a same level, from 53°20″ to 66°40″ the same portion is compared to the Head, in line with the same, the 6th group of Nakshtra spreading over in the cosmos between Head and Neck portion in descending order and ends at the 80°. Similarly, the 7th group which is also found to be in the descending order starts at 80° and reaches the 93°20″ at the Southern Cosmic Sphere which connects the Northern and southern hemi-spheres of the Cosmos. Similarly, the 8th and the 9th which are in continuation with the 6th are spread over in the Cosmic Southern sector in the descending Order. Following the ninth group the 10th 11th, 12th and the 13th groups of stars spread over in the Southern Cosmic sector in the same descending order. The 14th, which is spread over at the Southern tip of the cosmic sphere as the 5thGroup in the Northern Sector between173° 20″ and 186° 40″.

In continuation of the 14th group which spread at the Southern tip portion the 15th, 16th, 17th, 18th, 19th, 20th, 21st, 22nd, spread over in the Northwest Cosmic sphere in Ascending order from 186° 40″ to 293° 20″ consecutively. It seems that the 21st group is the connecting group between the Southern and Northern hemispheres of the cosmos as that of the seventh group of Nakshatra.

The 23rd, which is found spread-over in the Northwest portion of the cosmic sphere, is also in the same level and has no Ascending or descending order like the 5th group of Nakshatra which spread over in the North-east region of the Cosmic sphere. Following the 23rd group the 24th, 25th, 26th, and the 27thgroups linking with each other and spread-over in descending order between 306° 40″ and 360° and merges with the starting point of the 1st, group at 0° at the foot or parallel to the Cosmic equator level.

The Surya Siddhantha, an astronomical Text describes the shape of the Cosmos. The Cosmos is seen in-between the two bowls, which are inversely positioned each one at the Northern and southern tips. All the 27 Nakshatra groups are spread between the circles placed in the space between the two inverted bowls of the Cosmos. The Indian Almanac describes the ascending and descending positions of the 27 Nakshatra groups compares to the Human structure. The Krishna Yajur veda Sloga also describes the Cosmos. The Cosmos, which is covering by the 27 Nakshatra groups and within the hallow space and the position of the planets between the nakshatras and the Earth are Saturn, Jupiter, Mars, Sun, Venus, Mercury, and the Moon respectively, They compared it to the shape of an Egg or to a Pine-apple. All the above details made available only in the form of descriptions and in the form of Slogas, No graphical or pictorial views about the Cosmic-sphere, as detailed in the Astronomical texts. The Ancient Indians found that the different Nakshatra groups have their starting points in the same layer, which compared to the Human structure.

The below table shows the details:

S. No	Name of the Nakshtra group	Location in the Cosmic-Sphere from 1 to 27 (*Stars groups referred with reference to Indian almanac*)
1	1st Aswin., 9th Ayily, 10th Magha, 18th Ketta., 19th Moola and 27th revathi	All the six groups of Stellar start at the parallel and at the same level to the Cosmic Equator, in their locations in the North, North-east, South-west and North portion in ascending and descending order
2	2nd Bharani, 8th Poosam, 11th Pooram, 17th Anusha, 20th Poorada & 26th Uthirada	All these 6 groups begin at the first-layer (level) next to the Cosmic equator both in the North and in the Southern–hemi-sphere at Northeast, Southeast, and in South-west and North-west portions in ascending and descending order.
3	3rd Kritiga, 7th Punarvasu, 12th Uthiram, 16th Visaka, 21st Uthirada, 25th Poorattadhi	These stellar groups start from the 2nd level layer from the equator both north and southern hemi-spheres at North-east, East, South-east, South-west, West, and North-west portions of the Cosmic sphere in ascending and descending order.
4	4th Rohini, 6th Thiruvathirai, 13th Hastha, 15th Swathi, 22nd Thiruvona, 24th Sadayam	These groups of Nakshatras spread-over in the cosmos from the 3rd level layer from the equator in the both hemi-spheres in the Sorth-east, Southeast, South-west, North-West portions in the same manner of the other groups of stellar.
5	5th Mirugasheer, 19th Chitra 23rd Avitta	The 19th group stellar is positioned at tip of the Southern-hemisphere, the other two groups are positioned one each at North-east and North-west portions of the Cosmic-Sphere.

To identify the 27 various Nakshatra groups in the Cosmic-Sphere the Ancient Indians derived the Figurative system. In which all the Nakshatra groups are compared to the Worldly Living and Non-living matters. To assess the movement of the Sun which has its movement along and parallel to the Fixed and the integrated form of the nakshatras they combined more than two groups and formed 12 types of Constellations.

The following tables serve the details at a glance.

S. No	Name of the Nakshatra Group and their angular placement found in the Cosmos	Figurative assessment to identify the each stellar group in the Cosmos
1	Aswin 0° – 13°.20″	Spread over 0 to 13° 20″ tangential arch space in the ascending order and looks like a Horse -face The moon transits through in West-Eastern direction.
2	Bharani 13° 20″ – 26° 40″	As the same, this group spreads between 13° 20″ to 26° 40″ and looks like an Oven.
3	Kruthiga 26° 40″ – 40°	Occupies the space between 26° 40″ to 40° and looks likes a serious of lamps kept one over the other. The Moon travels in the South-North direction.
4	Rohini 40° – 53°20″	This group looks like a crown. The Moon moves through the North portion of the group
5	Mirugasheer 53°20″ – 66°40″	This group seems like the three holes in the coconut shell and spread over horizontally in the same level.
6	Arudhara 66°40″– 80°	This group formed out of a single star, which spread over in descending order.
7	Punarvasu 80°– 93°20″	This group formed mainly out of 7 stars over a base of 76 stars, which looks like the bottom of the Ship in the descending order. This group connects the northern and the Southern Hemi-spheres of the Cosmos.
8	Poosam 93°20″ – 106°40″	This group formed out of 4 stars and looks like a pumpkin flower inDescending order The Moon travels in the south portion of the group
9	Ayilya 106°40″ – 120°	This group seems to be in a rectangular. The Moon travels south portion of this.
10	Maha 120° – 133°20″	This is in the form of a plough. The Moon travels right across this group in ascending order from North-south.
11	Poora 133°20″ – 146°40″	This group proceeds in the same direction of the above group towards south direction and looks like the four legs of a cot.
12	Uthira 146°40″ – 160°	This group looks likes the same as the above group and the Moon transits towards south.
13	Hastha 160° – 173°20″	This group looks like the 5 fingers in the hand (or) in the form of a ear-ring.
14	Chitra 173°20″ – 186°40″	This group is at the South most bottoms of the cosmos in the same level parallel to the Cosmic equator

S. No	Name of the Nakshatra Group and their angular placement found in the Cosmos	Figurative assessment to identify the each stellar group in the Cosmos
15	Swathi 186° 40″ – 200°	This is a illuminant flame of a lamp and is the tail-end of the Saptharishi Stars in descending order towards the Cosmic equator.
16	Visaka 200° – 213 °20″	This group formed out of 7 stars and looks like a damaged constructed wall portion.
17	Anusha 213° 20″ – 226° 40″	This group formed by 9 stars and looks like the Lotus flower in the descending order. The moon moves right across this.
18	Ketta 226° 40″ – 240°	Shaped like a spear and the Moon travel south of this group in descending order/
19	Moola 240° – 253° 20″	This looks like a long mouth organ and in ascending order.
20	Poorada 253° 20″ – 266° 40″	This is a continuing group of the above in ascending order.
21	Uthirada 266° 40″ – 280°	Combination of the group this group looks like the legs of a cot. The moon travels north of the above group and right in the middle of this group.
22	Sravana 280° – 293° 20″	This is in the form of a yardstick in ascending order. The moon moves south of this.
23	Avitta 293° 20″ – 306° 40″	This group looks like a wheel and the moon travels south of this parallel to the cosmic equator.
24.	Sadaya 306° 40″ – 320°	This group contains 100 stars and looks like a bunch of flowers in descending order towards the cosmic equator. The moon travels south portion of this.
25	Poorattadhi 320° – 333° 20″	This is continuation of the above group in descending order.
26	Uthirattadhi 333° 20″ – 346° 40″	The combination of this group with the above group looks lik the 4 legs of a cot. The moon travels south among these groups.
27	Revathi 346° 40″ – 360° (0°)	This group looks like a fish, which is descending order, and joins where the 1st group starts parallel to the cosmic equator. The moon travels south of this.

THE CONSTELLATION

The Cosmos is formed out of the 27 Chain-linked Nakshatra/Stellar groups which are linked with each other and forms the hallow egg portion. After Noticing the movement of the Planet Moon which is positioned next to the Earth in the Cosmos, which completes one round along and parallel to the fixed and the integrated form of the Stellar groups in its position around the Earth at one stellar group/day which comes out 27 days.

But the Movement of the Planet Sun in its position takes 365 days to complete one round along and parallel to the fixed and the integrated form of the 27 Stellar-groups around the Earth in the Cosmos. The Planet Sun takes 28 or 29 or 30 or 31days to move 2 and a part of stellar groups. To define this ancient Indians bifurcate the 27 Stellar groups in to 12 groups and named them as constellations and thus the movement of the Sun along and parallel to the Constellations is about 365 Days or a Year.

Below Table shows the various constellation formed out of 27 stars.

S. No	Name of the constellation	The figurative assessment to identify the each constellation in the Cosmos
1	Mesha 0° – 30° (Aries)	This constellation is arrived by the combination of the 1st, 2nd, and a part of 3rd Nakshatra groups and looks like a Goat in the 0° – 30° ascending angular space from the Cosmic equator.
2	Rishaba 30° – 60° (Taurus)	In continuation of the above, the integrated part of remaining the 3rd, 4th, and middle portion of the 5thNakshtra groups compared to the shape of a Bull between 30°-60° angular spaces in ascending space above the cosmic equator.
3	Mithuna 60° – 90° (Gemini)	This Constellation starts from the mid portion of the 5th group and spread over in ascending order combined with the 6th and up to the 3/4th portion of the 7th group. The 7th group, which starts in the Northern hemi-sphere and passes through the cosmic-equator, ends in the Southern hemi-sphere, where the Sun transits through this group beyond 90° and termed as the starting of the Dakshina ayana of the Sun in the Cosmos. It looks like the two women standing nearby.

S. No	Name of the constellation	The figurative assessment to identify the each constellation in the Cosmos
4	Kadaga 90° – 120° (Cancer)	This constellation comprises of the 10th, 11th and 1/4th part of the 12th Nakshatra group from 90° – 120° in ascending order from the cosmic equator to the southern direction in the cosmos. This seems like a Crab.
5	Simmah 120° – 150° (Leo)	This constellation identified like the face of a lion. The 10th, 11th, and 1/4th of the 12th group of stars shape this constellation, which can see in the southern hemi-sphere of the Cosmos.
6	Kanni 150° – 180° (Virgo)	From the 1/4th of the 12th, 13th and up to the mid portion of the 14th star groups, the Virgo constellation is formed in the south of the cosmic sphere. It looks like a teen-age girl.
7	Thula 180° – 210° (Libra)	This constellation starting from the mid-portion of the 14th star group, which is at the dead end of the Southern cosmic sphere, which proceeds in descending order along the 15th and up to the 3/4th portion of the 16th group. It looks like a balance, which suspended with two-pans on either side of a rod with an indicator at the centre of the rod.
8	Virchiga 210° – 240° (Scorpio)	The Scorpio constellation stars from the 3/4th of the 16th star groups and continuing through the 17th and ends with the 18th Star groups. It is in the form of a Scorpion, in the southwest portion of the Cosmic southern-hemi sphere.
10	Magara 270° – 300° (Capricon)	This constellation starts from the cosmic equator region and extends through the 22nd and ends in the mid-portion of the 23rd group in ascending order in northwest hemi sphere of the Cosmos, It looks like a water-born living sea horse.
11	Kumba 300° – 330° (Aquirius)	Aquiris Constellation starts from the dead-end of the northern hemisphere spreads in the descending order towards the cosmic-equator along the 24th and up to the 3/4th portion of the 25th group in the northwest portion of the Cosmic-sphere, It looks like a ornate metallic Pot contained with purified water inside of the pot and covered with a coconut.
12	Meena 330° – 360° (Pisces)	Continuing the 25th group it reaches the Cosmic equator level through the 26th and the 27th group of stars it merges with the 1st group of stars which starts at 0° parallel to the Cosmic equator in descending order. It looks like a pair of fishes facing their tail ends with each other.

THE INDIAN CALENDAR SYSYTEM

The Ancient Indians found that the environmental changes around them on the Earth have the direct and the indirect influences of the cosmic bodies such as the sun, the moon and other planets and the Nakshatras groups around the Earth. After a keen observation over the systematic changes in the climate, ancient Indians located the position of the Stars, Sun, Moon and other planets in the Cosmos. It can be well understood beyond any doubt from the Almanac designed by the Ancient Indians and the Astronomical text The Surya Siddhantha, and other related texts that the Ancient Indians were the pilots not only in discovering the Astronomical facts but also in discovering the Earth as a spinning sphere.

To locate the positions of the planets in the Sphere of the Earth, they bi-sected the sphere with a line at the center and termed it as *Bhu-madhiya rega, the axis of the earth sphere as Bhu-achu or otherwise known as Cosmic-equator and cosmic pole-line.* To have more accuracy, further they sub-divided the sphere area by introducing the Atcha and Deerga rega, later, which was known as Longitudes and Latitudes. They were well aware of the Sunrise and sunset at varies places and gradually differs due to the spinning movement of the Earth in the Cosmos. They used to ascertain the local mean-time based on the actual moment of the Sun. Through the shadows formed on the earth, by the Sun rays they found more methods to arrive the actual mean time.

8. 1. ASTROLOGY AND ASTRONOMY

Astrological theories have been evolved on the basis of the Ancient Indians Earth centred cosmic principles. In Earth centred cosmic principles all the planets which are in 8 kinds of motions have their movements with reference to the fixed stellar-groups around the Earth in the Cosmos. The planets in the cosmos acts like lenses and converge the energies released from the stellar groups within the hallow space of the cosmos which is known as Maha-praanaa, these energies will be continuously released over the surface of the earth. Thus these Maha-praanaas from the stars and the Planets are responsible and major cause for the Creation, conservation and ruination activities which happens across the Earth. The Indian Vedaas are the Part and Parcel of the Astronomy and Astrology which deals with the study of Time-equation known as Skandharaimaham which contains three skandhas. Aaryabhatiyam by Aryabhatta, Varaha hora by Varahamihira, Parama siddhantha by Premavallba, Hora Sastra by Bharat Banesh, Bharat Yuvana Jathagam, Surya Siddhantha by Kabileshwar Sastri, Guru Vichara, Sani Vichara, Vimanathala vakara Vichara (Analysis 0d retrogression of planets in the Cosmos), Chandra Vichara by Dhanyanantha which are some among the various texts to analyse the Astronomy through Astrology.

These Ancient Indian Astronomers with their observation over the space, they noted the gradual development and the gradual depreciation of the Moon in space at some frequent intervals also they noticed that particular stars groups are visible in the space for a long-duration. They examined the visibility of the star groups thoroughly and they separated the planets Mercury, Venus, Mars, Jupiter, and Saturn from the Stars groups while they too seem like the stars in the Space. To arrive the exact locations of the Star-Groups, Planets position from their observation points, they evolved a detailed time calculation method in which they used the time-duration of inhaling and exhaling of normal human breath and they have considered the time duration between the opening and shutting of the eye-lids of humans, and they considered the same as the basic units for time.

Time-calculations:

1 swasa (breathing time)	1 Prana (Normal time between inhaling and exhaling)
6 prana	1 palams
60 palams	1 nashigai
60 nazhigaies	1 Day or (One Arohana or oneAvarohana Nakshatra through which the Planet moon transits)
15 Days	1 paksha (Which denotes the movement of the Moon along and parallel to the fixed star-groups of the fixed star-groups in the Cosmos either in developing or in diminishing in her shape)
2 Pakshas	1 month (which denotes the transit of the Sun along the Parallel to a constellation)
12 months	1Year (or) 1 Deva Day (which denotes the time taken by the Sun to Complete one round along and parallel to the Star-groups fixed in the Cosmos with reference to the earth)
360 Deva days	1 Deva Year

1 chatur yuga	1 Manvanthara (Chatur Yuga consists of four Yugas. 1. Kali. 2. Duvapara. 3. Thretha and 4. Krutha Yuga
14 Manvantharas	1 Kalpa (There are 7 kalpas in practice and the presentRunning Kalpa known as Sriswedavaraga Kalpa which is found to be first one among the other 6 Kalpas).

In Indian Astrology, the KAALAPURUSH which has been referred to the Anatomy of an Average man on the Earth, to understand the fixed and the integrated position of the stellar groups in the Cosmo: the Cosmos is divided into two hemispheres namely the Northern and the Southern Portions. The equator's extended portion in the Cosmos is termed as Cosmic Equator and the portion of the Earth's Pole-line termed as cosmic pole-line. According to the human anatomy, a man's height is measured from 8 parts from his/her little finger-tip to thumb finger tip. Using the above fact, the Ancient Indians compared the Cosmos to a normal humans Anatomy and distributed the starting points of the 27 stellar groups around the Earth. They allotted One Kaalapurush's feet rest on the Cosmic Equator and another's feet rest on the same cosmic equator and stands on both North and Southern Hemi-spheres.

The average age of a man on the Earth is found to be 120 years. It is considered that the influence of Each Planet on an average human starts from the Moon's transit position over a particular stellar groups. All the planets within the cosmos which are revolving around the Earth have different span of periods as follows:

Planets

Planets	Span
Kethu (shadow Planet)	7 years
Venus	20 years
Sun	6 years
Moon	10 years
Mars	7 years
Raahu (shadow Planet)	18 years
Jupiter	16 years
Saturn	19 years
Mercury	17 years

The starting point of 27 stellar groups in arohana (ascending) & avarohana (descending) orders in the cosmos with reference to kaala purush as follows

25

Kala purush feet level;

- Kethu (feet)   Aswini, Maham, Moolam
- Venus (thigh)   Bharani, Pooram, Pooradam
- Sun (Stomach)   karthiga, Uthiram, Uthiradam
- Moon (Neck)   Rohini, Hastam, Thiruvonam
- Mars (Head)   Mirughaseer, Chitra, Avitam
- Raahu (Neck)   Thiruvathara, Swathi, Sathayam
- Jupite (Stomach)   Punarvasu, Vishaga, Poorathadhi
- Saturn (thigh)   Poosam, Anusham, Uthirattadhi
- Mercury (feet)   Ayilya, ketta, Revathi.

One must have the basic understanding about the Indian Cosmology, the Earth Centred Cosmic theories including the Astrological theories, then it is obvious that the twinkling in the Space are positioned in multiple layers of circular orbits and the planets are oscillating in their positions in South-North and North-South directions based upon the Cosmic-Equator both in Northern and Southern hemi-spheres of the Cosmos. In General Indians have faith in the Astronomy, For instance, based on the Scientific Astronomical theories from the ancient Indians, they used to go to the Astrologers to fix the matching among the couple, most of the Astrologers knows the values of the Raahu among the ten major-points of the suitability of the birth-charts of the couple. In the Indian Cosmological theories, the RAAHU referred to the starting point of the 27 stellar-groups in the Cosmos. The term Arohana/Ascending and the Avarohana/Descending positions of the stellar groups in the Cosmos with reference to the Cosmic Equator and Cosmic-Pole-line. Using ascending & descending order of stellar groups, parts of the kaala purush & the dasa periods of the planets in the astrological theories with the Indian cosmic principles this Indian cosmography have been arrived in the following manner.

1. The periods (Dasabukthi) of the planets are taken as the diameter of the circles.

2. Cosmic Equator: This the horizontal line extended from the earth's equator on the both sides of the earth & as the feet part of the kaalapurush. This line or layer is common for both northern & southern hemisphere (1cm=1year)

3. Cosmic Pole: This is the line extended from earth pole line on both sides of north & south portions of the earth.

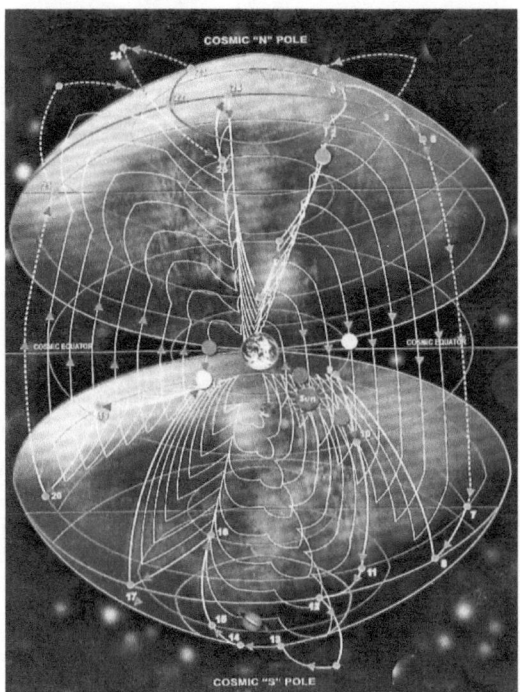

Indian Cosmograph

4. Common Layer:(The Feet portion of the Kaalapurush) At this potion, the Dasa Periods of the Planets Mercury and Kethu is represented with two circles of dia 17&7cm are drawn with the Earth's centre and the starting and ending points of the Stellar groups (1) Aswin, (10) Maham, (19) Moolam,(9) Ayilyam, (18) Kettai, (27) Revathi are marked in angular directions.

5. 1st Layer in Northern Zone: This layer fixes at the thigh level of the Kaalapurush and at 4 parts from the feet of the Kaalapurush in which the Dasa periods of the Planets Venus and Saturn 20 years and 19 years respectively and the 2 circles of dia 20cm and 19 cm have been drawn on pole line as the Centre and star groups (2) Bharani and (26) Uthirattathi are ploted with their angles from the dia. of the circles which are parallel to the Cosmic Equator.

6. 2nd Layer in Northern Zone: At 5 parts of distance (6. 25cm) on the pole-line, two circles of Dia. 16 & 6 cm are drawn and the star groups (3) Karthigai, (21) Uthiradam, (7) Punarpoosam, and (25) Poorattadhi are fixed with their respective angles.

7. 3rd Layer in Northern Zone: At 7parts of distance (8. 75cm) on the cosmic pole-line from the centre of the equator, 2 circles of Dia. (18&10 cm) are drawn and the star groups (4) Rohini, (22) Thirvonam. (6) Thiruvathirai, and (24) Sathayam are plotted in their respective angular portions.

8. 4th Layer: At 8 parts of distance (10cm) a 7 cm Dia. circle has been drawn pole-line as its centre and the stellar groups (5) Mirghasheer and the (23) avittam are plotted in their respective angular portions.

    Southern Zone: All the circles are drawn with the centres on the cosmic pole like in the northern zone in descending order and representing the following stars with their respective angles.

    1st Layer: Pooram (11), Pooradam (20), poosam (8) and Annsham (17)

    2nd Layer: Uthiram (12), Uthiradam (21), punarpusam (7), Visakam (16)

    3rd Layer: Hastham (13) and Swathi (15)

    4th Layer: Chittirai (14)

9. Finally all the longitudinal points of each stars has been connected with arched tangents continually the area of which all the star groups are regulated.

10. Through this Cosmograph, thus arrived is sufficient to prove the basic concepts of the earlier valuable theories of Indian Astronomy.

11. Further studies can give the exact distances in Yojanas or in Kilometres as desired.

12. If we go through the plan of this cosmography, it can be found that all the stars are positioned in a lotus-petaline from which the basic concept of the Indian Yoga and Tantra systems can be well understood.

13. All the stars which have been imagined are identified with the figures of worldly things. Though the imagined figures of worldly things of stars have not been illustrated in the cosmography the same may be arrived by imagining such figures on the basis of the cosmograph.

14. The authenticity of this cosmography can be verified by analysing the movement of the sun with reference to the fixed and integrated form of the stellar in the cosmos. The star Dhakshanayana occurs when the sun enters through the stellar group (7) punarpusam at its fourth quarter, similarly the Utrayana starts when the sun enters through the second part of the 21st stellar group Uthiradam.

8. 2. THE ANCIENT INDIANS' CALENDER SYSTEM

The Ancient Indians were well aware, that the normal life of the human beings on the Earth will not be smooth without any proper planning. Hence they invented their calendar system based upon their Geo-centric and the Oneness Universe Cosmological Theories. They identified the impacts of the 27 stellar, 7 planets, and the shadow planets over the Earth and identified the Fixed and the integrated form of the 27 stellar-groups around the Earth in the Cosmos. The Stellar energy is the basic source for the movement of all the planets, which have their movement along and parallel to the stellar energy-line which is in the form of PETAL-LINE PATH, This PETAL-LINE PATH, will be either Circular nor elliptical orbits. They considered all the Planets and the fixed positions of the stellar as the basic factor in arriving their detailed Calendar-system.

8. 3. HORA-SYSTEM

It is learnt that the term "HORA" MEANS the duration of time of a particular planet's influence over the Earth. Accordingly, the Ancient Indians identified the duration of time between the two consecutive Sun raises as 60 NAZHIES and later which is known as 24 HOURS.

8. 4. ORDER OF THE WEEK-DAYS

The Ancient Indians noticed the routine of Sunrise in the cosmos, from which they found-out a way to identify the 60 NAHIES (24 Hour) limit by using the order of the planets which are positioned in the Geo-cantered cosmos and their influence over the Earth. From their Slogas we can identify the order of the planets from the stellar are as follows: 1. Saturn, 2. Jupiter, 3. Mars, 4. Sun, 5. Venus, 6. Mercury, and 7. Moon. They distributed an even time-bound influence for each planet in the above order at the rate of 2–1/2 Nahies or one hour between the two consecutive sunrises.

Basis the order of the planets, between the two consecutive sunrises there are around three complete rotation which ends at 21 hours, Thus 3 complete cycles of the order of the planets merges with the Sunrise, when 21 hours lapsed the 4th turn starts with the 1st planet Saturn and the 24th hour ends with the planet Mars. Following this the 25th hour which merges with the new sunrise which co-inside with the next order of the planet Sun. Continuing this cyclic pattern with reference to the order of the planets in the cosmos the third sunrise starts with the planet Mars, the 4ᵗʰ sunrise with the Mercury, the fifth sunrise merges with Jupiter and the 6ᵗʰ Sunrise starts with the Venus. Using the above rotation Ancient Indians arrived at the order of the weekdays AS SATURDAY, SUN-DAY, MON-DAY, TUES-DAY, WEDNESDAY-DAY, THURS-DAY, AND FRIDAY.

FIVE-ELEMENTS FACTOR

The Ancient Indians gave more importance to the Five-Elements Factors in designing their Calendar system. The basic and primary five elements are 1. The Air, 2. The Heat, 3. The Water, 4. The Earth Energy i. e., the Gravitational force, 5. The energy received from the Stellar and the Planets in the cosmos around the Earth.

DAY

A Day is defined as the time spread between the two-consecutive Sunrises and identified with the order of the planets, which are positioned between the integrated form of stellar and the Earth. Initially the first Sunrise is identified as Saturday as the Saturn occupies the first position from the Stellar and the subsequent sunrises are identified as Sunday, Monday, Tuesday, Wednesday, Thursday, and Friday from the order of the planets Saturn, Jupiter, Mars, Sun, Venus, Mercury, and Moon between the Stellar and the Earth.

NAKSHTRA OR STELLAR GROUP

The influence of a particular Stellar Group in the Cosmos can be identified on a particular day by using the movement of the Moon through the stellar group.

THITHI OR FULL/NEW-MOON

The Moon's movement has been compared with the position of the Sun's movement in the Cosmos with reference to the fixed stellar constellation in the cosmos. For Example, in the month of April the Sun is used to be in transit along and parallel to the Stellar groups namely Aswin, Bharani and krithika's first quarter portion. During that period when the moon takes it transit through the above stellar groups in its position, the Sun and moon happens to be at the same stellar groups in their positions which represents the No moon or otherwise known as Amavasai. As the Moon has a lesser circumference petal-line orbit than the Sun, it has to move along the stellar force-line and reach the opposite phase where the Sun has a lesser transit when compared to the transit of moon. Now the Moon has its transit within the stellar groups, and the Sun and the Moon happens to be at a straight-line and this forms a Full moon which is known as Poornima.

It is quite interesting to note that the different phase of the moon takes place due to its movement along and parallel to the curved placement of the stellar groups in the cosmos and their force-line with reference to the cosmic equator.

YOGAS AND KARNAS

The Ancient Indians made their interpretations of the Yogas and karnas based upon the positions of the moon, the Sun, the Stellar, and the other planets influences over the earth on a particular day.

THE CALENDER

The Ancient Indians derived their Calendar system using the Nazhigai as the basic unit of time measurement. The following table will provide the details of the Ancient Indians Calendar system.

An eye-lid movement = time taken to produce a sound with the thumb and the middle finger (Nodi)

2 Nodies = 1 Mathirai

2 Mathiraies = 1 Guru

6 Gurus = 1 Swasa or Prana

6 Pranas = 1Kshna or palam

12 Kshnas = 1 Vinazhi

60 Vinazhies = 1 Nazhi

2. 5 Nazhies = 1 Hour

60 Nazhies = 24 Hours or 1 Day

30 Days = 1 Month (The moon completes one round along and parallel to the fixed stellar-constellations in the cosmos.)

12 Months = 1 Year (The Sun completes a one round along and parallel to the fixed stellar constellations in the cosmos.)

SIXTY YEARS OF CYCLE

The Ancient Indians further expands the year system to have greater analysis for the future forecast using the five-elements. They sub-divided the sixty-years (complete details on sixty years can be refered from the annexure) in five major heads such as 1. ETHAVATHSARA, 2. EDAVATHSARA, 3. SAMVATHSARA, 4. PARIVATHSARA, and 5. ANUVATHSARA. Using the VATHSARAS they used to forecast the Climatic changes and other important occasions that would happen on the Earth.

AYANAS

Ayana means movement. The Ancient Indians derived the Ayanas based upon the movement of the Sun in the Cosmos. They identified the movement of the Sun through and parallel to the Fixed Stellar constellations which are positioned the in Northern hemi-sphere of the Cosmos is known as Uthira-Ayana and the movement of the Sun through the stellar constellations in the Southern hemi-sphere of the Cosmos is known as DASHINA-AYANA. Normally when the Sun enters through the stellar constellations in the Northern-hemi-sphere during the middle of January and transit in the Northern portion of the Cosmos until the middle of July of every year and from the middle of July and to the middle of January of every year the Sun happens to be at the Southern hemi-sphere of the Cosmos.

The Ancient Indians elaborated their calendar system as follows:

12 Months	1 Deva Day
360 Deva Days	1 Deva Year
1200 Deva Years	1 Chathur Youga*
71 Chathur Yougas	1 Manvanthra**
14 Manvanthras	1 Kalpa***

* Chathur means four, the four Yougas are 1. KIRUTHA YOUGA, 2. THREYDHA YOUGA, 3. DHUVAPARA YOUGA and 4. KALI YOUGA.

The various time-limits for the four Yougas are as follows:

1. KIRUTHA YOUGA= 4800 17, 28, 000
2. THREYTHA YOUGA= 3600 12, 96, 000
3. DHUVAPARA YOUGA= 2400 8, 64, 000
4. KALI YOUGA= 1200 4, 32, 000

NAMES OF THE FOURTEEN MANVANTHARAS:

1. SWAYAMBU MANVANTHARA
2. SWAROSHTA MANVANTHARA
3. UTHAMA MANVANTHARA
4. DHAMASA MANVANTHARA
5. RAIVATHI MANVANTHARA

6. SAKSHEESHA MANVANTHARA

7. VAIVASUTHA MANVANTHARA

8. SOORYASAVARNI MANVANTHARA

9. DAKSHINASARVANI MANVANTHARA

10. BHARAMASAVARNI MANVANTHARA

11. DHAMASAVARNI MANVANTHARA

12. RUDHARASAVARNI MANVANTHARA

13. DEVASAVARNI MANVANTHARA

14. INDRASAVARNI MANVANTHARA

KALPAS

There are seven kalpas which are known as:

1. SUVEDHAVARAHA KALPA

2. KOORMA KALPA

3. PARTHIVA KALPA

4. SAVITHRI KALPA

5. PRALAYA KALPA

6. VARAHA KALPA

7. BRHAMA KALPA

ARITHMETIC CALCULATIONS

The Ancient Indians used 36 digit numerals to have an elaborate astronomical calculation and they named them according to their placement. They are as follows: 1. First digit as one, 2. second digit as Dasamam, 3. Third digit as Sadam (hundred), 6th digit as one lack, 7th as ten lakh, 8th as one crore, 9th as dasa-kodi, 10th as sadha-kodi, 11th as Arbhudham, 12th as Nirbhutham, 13th as Garvam, 14th as Maha-garvam, 15th as Padmam, 16th as Maha-padmam, 17 as Kshoroni, 18th as Maha-kshoroni, 19th as Sangam, 20th as Maha-sangam, 21st as Kshathi, 22nd as Maha-kshathi, 23rd as Shodam, 24th as Maha-shodam, 25th as Nidhi, 26th as Maha-nidhi, 27th as Bharatham, 28th as Parthin, 29th as Anandham, 30th as Saharam, 31st as Aryam, 35th as Bhoori, and the 36th digit is known as Maha-bhoori. The detailed formation of the Indian Calendar System is standing as a proof for their Oneness Geo-centric cosmic theories.

9. 1 IDENTIFICATION OF THE ORDER OF THE DAYS IN A WEEK

The Ancient Indians defined the period of 60 Nazhigaies between a Sunrise and the Sunset, to identify the same they used the order of the planets between the 27 Nakshatra groups and the Earth in the Geo-centric principles. **The orders of the planets in the Geo-centric theories are 1. Saturn, 2. Jupiter, 3. Mars, 4. Sun, 5. Venus, 6. Mercury and The Moon from the stellar respectively.** Considering the starting of a new yuga, the 1st positioned planet merges with the initial sunrise of that yuga for a period of 2 1/2 Nazhigaies (or) an hour. Continuing the Saturn, the other planets merge with the Sunrise in a cyclic manner. Thus 3 complete cycles of the order of the planets merges with the Sunrise, when 21 hours lapsed the 4th turn starts with the 1st planet Saturn and the 24th hour ends with the planet Mars. Following this the 25th hour which merges with the new sunrise which co-inside with the next order of the planet Sun. Continuing this cyclic pattern with reference to the order of the planets in the cosmos the third sunrise starts with the planet Mars, the 4th sunrise with the Mercury, the fifth sunrise merges with Jupiter and the 6th Sunrise starts with the Venus.

The adoption of Planets between the sunrises, and the arrival of the weekdays the system known as the "Hora" and Prof. H. H. Wilson expressed that the origin of the arrangement not ascertained very precisely, as it was unknown to the Greeks and not adopted by the Roman until a late period. It commonly ascribed to the Egyptians, but no sufficient authority and the Hindus in India to have atleast as good as to the invention of any other people. The Hora system, derived by the Ancient Indians is still in force and followed by the Indians in their daily occupations.

9. 2 IDENTIFICATION OF THE MONTH

The Ancient Indians identified the twelve cyclic months with reference to the transit position of the Sun and Moon along and parallel to the Nakshatras groups in their positions. Accordingly, they named after the Sun while it transits through twelve constellation namely 1. Mesha (Aeries), 2. Rishaba (Taurus), 3. Mithuna (Gemini), 4. Kadaga (Cancer), 5. Simah (Leo), 6. Kanya (Virgo), 7. Tula (Libra). 8. Virchiga (Scorpio), 9. Dahus (Sagittarius), 10. Magara (Capricorn), 11. Kumba (Acquires), 12. Meena (Pisces). In addition to that they identified months with reference to the transit position of the Moon through a particular Nakshatra (stellar) group when the Full Moon occurs. 1. Chaitram. (14th Group), 2. Vaisakam (16th group), 3. Jeyshtam (18th group), 4. Ashadam (20th group), 5. Sravanam (22nd group, 6. Pathrapadam (25th group), 7. Aajveejam (1st group), 8. kruthigam (3rd group), 9. Arudhara (6th group), 10. Pushiyam (8th group), 11. Magam (10th group), 12. Palguanm (12th group).

ORDER AND MOVEMENT OF THE PLANETS

The Ancient Indians found that the energies which are formed due to three factors like Air, Water, and Heat, which are the common factors over the Earth. Further, they compared the energies to the Stars, the Planets and classified as the cosmic energy. They invented that the five energies are main factors in deciding the environmental behaviour of the Earth. From constant and keen observations with these five factors of energies, they evolved 60 different years and identified them by individual names. They derived necessary formulae to ascertain the movement of the each planet with reference to the fixed and the integrated form of the 27 nakshatras in the cosmos for every individual year. From the above circumstance, it is easy to understand that at the time of formatting the Calendar system there was no religion. However, in later point of time, the followers of the geo-centric principles who had practiced made them as a religion. But in the recent centuries, the geo-centric cosmic theories which are treated as a part of the religious belief and removed from the consideration in scientific point of view, being it is assumed as based upon religious faith.

From the above Ancient Indians timetable, this can easily ascertained that the Studies about Cosmology in India took place even before 6 kalpas of period. From this, it can be realized initially the Ancient Indians had discovered the fixed and the integrated form of the 27 Nakshatras, the 7 kinds of Planets, the shape of the earth as a Sphere. Further they, based upon their discoveries made in the Cosmic Studies, they identified the shape of the Cosmic – sphere as an Egg or of a Pineapple.

In-side of the hallow egg-sphere, they positioned the Planets in the space between the Nakshatra groups and to the Earth in descending order (Saturn, Jupiter, Mars, Sun, Venus, Mercury, and the Moon). Since for the last few centuries, it is found that the followers of the geo-centric cosmic principles in India had forgotten the original, the fixed, and the integrated form of the 27 nakshatra and the movements of the planets along and parallel to the these nakshatras groups forms through which the entire cosmos is made. They simply believed only in the positions and order of the planets and their time-taken to complete one rotation in the cosmos around the Earth. They were unable to reconcile the fixed and the integrated form of the 27 nakshatras with reference to the earth, which are available in the Surya siddhantha and in the Almanac in a hidden manner.

The Ancient Indians, Chinese, Babylonians, Arabs, and other Western thinkers had formulated their findings and thus various treatises are available on this. However, the experts in the field like Cole-brook and Ebenezer Burgess are of the views that the Indians were the pioneers and for most on expounding on the subject. Burgess felt that the positions of the constellations, which could not precisely understood in the Hindu Astronomy by the then Indians and further studies of the Hindu astronomical concepts, could reveal more information and lead to have more enlighten. Further Burgess felt that the Hindu system of astronomy had relegated the positioning of the Stellar constellation to a lesser important place, the fact remains in the Hindu life style.

The Ancient Indians after discovering the basic factors such as the fixed and the integrated format of the 27 Nakshata groups around the Earth, the exact locations of the Planets between the Nakshatras they concentrated on the movements of the planets with reference to the Fixed stellar in the Cosmos. They had precisely ascertained that the cosmic energy is the basic cause for the movements of the planets with reference to the Fixed and the integrated form of the stellar groups in the cosmos and the Earth to have a stationery position at the center of the Cosmos.

The spinning movement to the earth is due to the excess energy realized at the center of the cosmos. Out of the two stellar groups, which covers the sides of the Earth in the Cosmos position, where the Earth positioned in the center of the Cosmos has an undisturbed regular spinning movement. Further, they understood the cause for the movement of the Planets along and parallel to the fixed and the integrated format of the stellar groups is the eccentric placement in the cosmic sphere, they critically investigated and found that positive as well as negative kind of forces from the stellar groups causes the movement of the planets, which are in eight kinds. The eight kinds of motions of the planets along and parallel to the fixed stellar groups in the Cosmos are 1. Vakra (Retrograde) 2. Anu-vakra (somewhat retrograde) 3. Kutila (Traverse), 4. Mandha (slow), 5. Mandhatara (Very Slow), 6. Sama (Even), 7. Cigara (Fast), and 8. Cigaratara (Very Fast).

They found that the Earth, which is at the center of the Cosmic-shell, has no movement along and parallel to the fixed position of the stellar groups like the other planets. The earth has the self-rotation movement, which causes the Gravitation force around the Earth in the Cosmic-shell. The other planets, completes one round starting from the parallel level to the cosmic equator and passes through in ascending as well as in descending order all over the inner shell of the cosmos, covering the entire northern and southern hemi-sphere and will finish one round from where they started.

They fully understood when the objects have their movements in ascending and in descending order with reference to the Cosmic equator, then the pathway so created by the stellar in the cosmos must be in an irregular position and compared the same to the Lotus Petal-line format.

The followers of the geo-centrism, had forgotten this critical pathway of the planets in the cosmos, No planet in the cosmos keeps a constant distance from the Earth, as they are in continuous movement along and parallel to the Fixed star groups behind them. So they provided the total circumferential distances for each planet in their positions and the total circumferential distance of the Fixed and the integrated form of the 27 stellar groups. In addition to that, they provided the total circumferential stellar energy extended beyond the stellar groups, through which it understood that the stellar energy could also spread all over within the cosmic-shell.

The following table provides the circumferential lengths.

Sl. No	Details	Circumferential length (in Yojanas)
1	The Universe	18, 112, 080, 864, 000, 000
2	The fixed and the integrated form of stellar groups	259, 890, 000
3	Saturn	127, 688, 285
4	Jupiter	51, 375, 764
5	Mars	8, 146, 909
6	Sun	4, 351, 000
7	Venus	2, 664, 637
8	Mercury	1, 043, 209
9	Moon	324, 000

Note: In the above table, the Yojana mentioned referred to the linear Astronomical Unit for which no exact equalling modern measuring unit derived. At present One Yojana is equated to 5 Miles or 8. 045 Kilometer (approximation).

The Ancient Indians, after arriving the total circumferential lengths of each planet and the fixed and the integrated form of the stellar groups and the energy line extended beyond outside of the stellar groups in the Cosmos, they discovered the average time taken by the each planet to complete one round along and parallel to the Nakshatra groups. Later, the then Investigators and Astronomers considered the movements of the planets along and parallel to the fixed stellar format as the Sidereal periods of the planets in the cosmos.

The following table provides the time-taken by the each planet to complete one round in the inner shell of the cosmos, along and parallel to the fixed stellar groups.

Sl. No	Name of the Planet	Time-taken to complete one-round around the Cosmos along and parallel to the Fixed and the integrated form of the Nakshatras
1	Saturn	30 years or 10956 days
2	Jupiter	12 years or 4383 days.
3	Mars	18 Months or 687 days
4	Sun	12 Months or 365 days
5	Venus	225 days
6	Mercury	88 days
7	Moon	27days.

It is noticed that when equating the data's shown in the above tables with the relation between the time and the distance that the Planets are moving in a equal speed of motion along and parallel to the fixed and the integrated form of the Nakshatra groups.

The following table provides the speed of the individual planet in the cosmos

Sl. No	Name of the Planets	Circumferential Length	Time taken to complete one Round (in Days)	Speed of the Planet (in Yojana/Hour)
1	Saturn	127, 688, 285	10956	485. 6083
2	Jupiter	51, 375, 754	4383	486. 3958
3	Mars	8, 146, 909	687	494. 1125
4	Sun	4, 351, 000	365	496. 3500
5	Venus	2, 664, 637	225	493, 4458
6	Mercury	1, 043, 209	88	493. 9416
7	Moon	324, 000	27	493. 9600

From the Speed of the Planets, it is understandable that all the planets in the Cosmos are in more or less in the equal speed of motion. With reference to the Indians Almanac, it is found that the planets, the Sun and the Moon only have not suffered by the retrograde motion among all others are facing the retrograde movements due to the Negative magnitude forces of the Nakshatras groups. This may be the cause to the Ancient Indians to give more importance to the Sun and Moon.

According to the Ancient Indians the Fixed and the integrated form of the Nakshtra Groups, which forms the Cosmic Shell, found not to be either in a regular Circular or in a regular elliptical shape but in a irregular format. This causes the movements of the planets in their position in the inner portion of the cosmic shell parallel to the Nakshatra groups in a irregular Shape, which looks like the outer circumference of the peals of a Lotus flower which is well described in the figure.

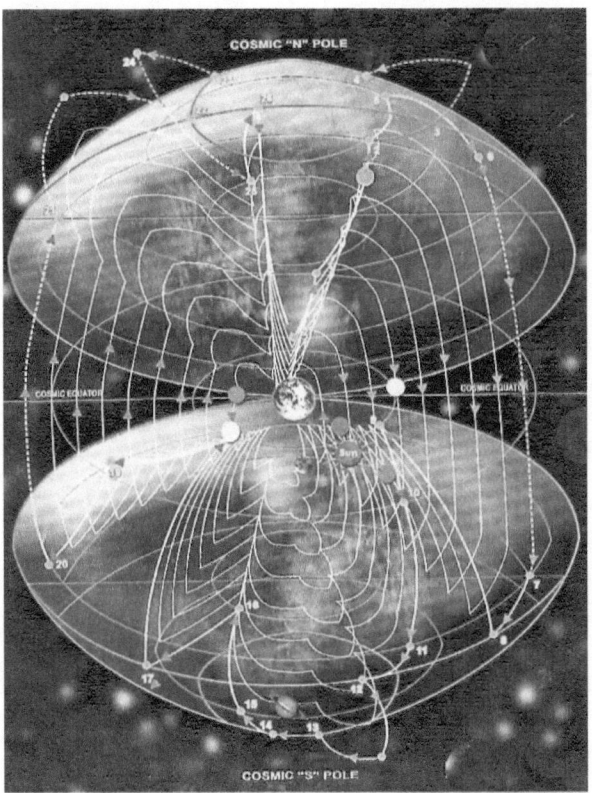

The above picture shows the planetary positions in the stellar covered cosmos as on 20. 08. 1953. At Thanjavur, Tamilnadu.

The below plan shows the fixed and the integrated form of the 27 stars in the cosmos with reference to the earth's equator and its pole line which looks like a lotus petal line pathway.

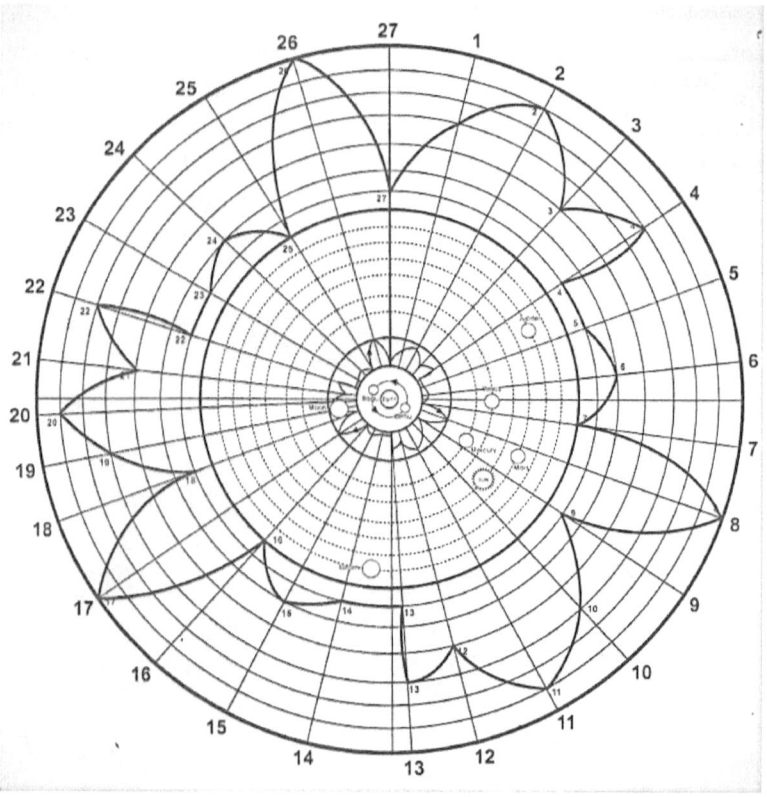

FORMATION OF CRESCENT AND PHASES OF THE MOON

The Ancient Indians *found that the Moon* takes an average duration of 27 days to complete one round through the lotus petal-line pathway and completes all the 360°. During the same period, the Sun transits in the petal-line pat way in its position only up to 30°. The commencement of the *new moon* takes place at any time when the moon transits between 0° to 30° tangential curvature space inside of the cosmos in its position and merge with the level and position of Sun in the same 0° to 30° tangential curvature space in its position. The angular deviation, the depart movement in the descending and ascending order of the moon with reference to the positions of the Sun, causes the Formation and the development of the Crescent Moon. When the moon reaches the opposite position to the Sun, *the Full moon* shines in the Cosmos. Continuing her movement in the cosmos towards the transit position of the Sun, which once again makes the moon merge with parallel level of the sun in her position in the next 30° to 60° Cosmic sphere and a *new moon* commences once again! The sufficient details provided in the Almanac about the day-to-day movements of all the planets with reference to the Fixed Nakshstras in the Cosmos.

The following Graph in the following page shows the details of the developing and diminishing shape of the Moon.

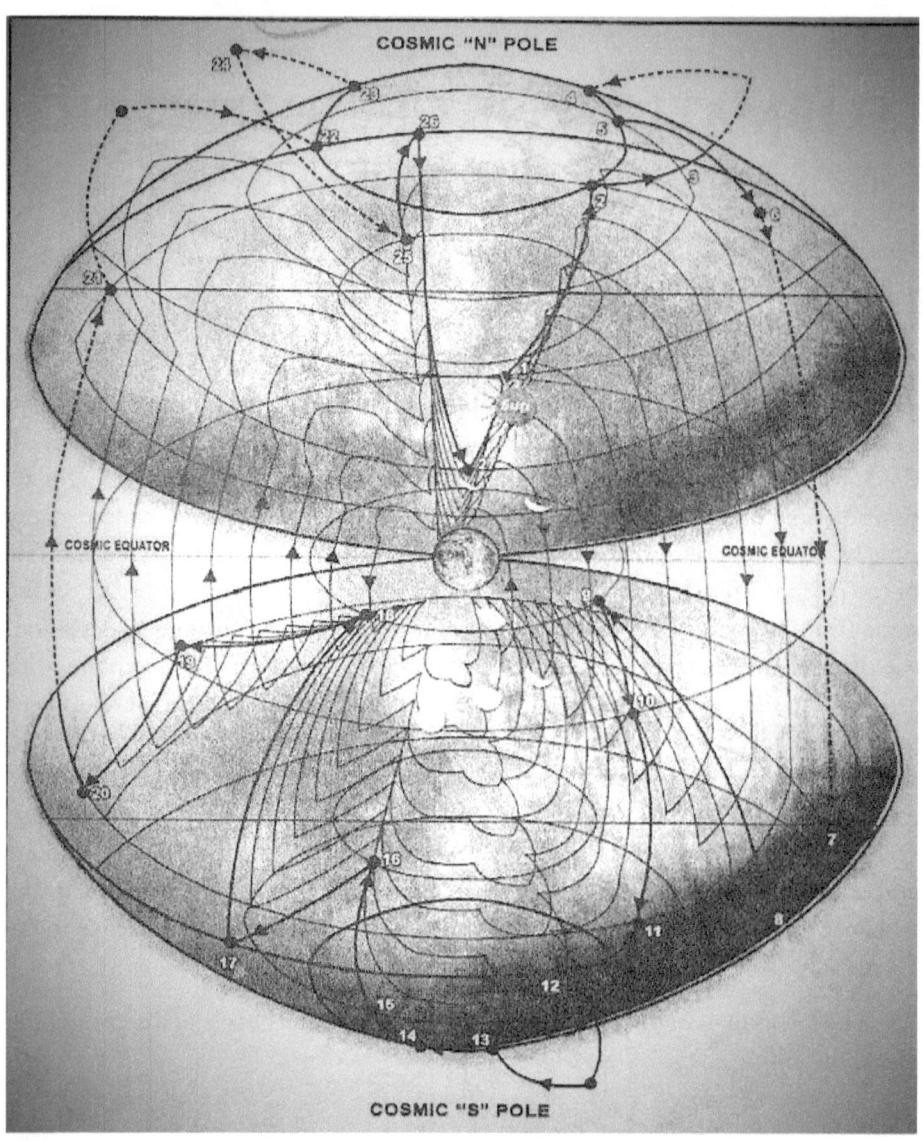

FORMATION OF THE LUNAR AND SOLAR ECLIPSE

The inventions and the critical studies about the formations of the lunar and the solar eclipses made by the Ancient Indians are the examples for their outstanding excellence of intelligence over the science of Physics! They knew when a bright object seen through a color screen then the object would also see as the color of the Screen. They undertook a closer observation and noted the color variation over the lunar and solar eclipses and both of them were seen light red and black in color. This made them that shade as screen, which falls over the Earth and obscure the bright objects the Sun and Moon while they are passing through it in the cosmos.

Then they invented the cause of the formation of the screen like shade and the movement of the same in the Cosmos. It is normal to see that the shade of an object have its direction of movement along the movement and direction of the object. They invented that the shade of an object, which has its shade casts through the penetrated rays of light behind it, which have its movement of direction, is in opposite direction of the object. They realized that the shades of the planets Saturn, Jupiter and Mars are formed due to the penetrated rays of Nakshatra groups behind them and identified them by their colors that are in Black, Red and Yellow respectively in the Cosmos.

From this, it is understood that if the planets have their movement either in a regular circular or in a elliptical orbit, then there is no possibility of their shades casts in the cosmos over the Earth to have an opposite direction of movement against the parent object movement. In the Lotus petal line movement of the Planets along and parallel to the fixed and the integrated form of the Nakshatra groups have the chance to have the opposite direction movement of their shades.

They omitted the other Planets shades in the cosmos as if, they are not formed over the Earth as the others. The Indian Almanac designed by the Ancient Indians provides a foreseen note on the formation of different types of eclipses, the duration, the color of the eclipsed planet for the 60 cyclic years!

It is found from the Indian Almanac that two pairs of eclipses will be formed during every year and a maximum of three pairs of eclipses will be possible once in 18 months due to the transits of the shade planets in the cosmos. Not all the eclipses forms in the cosmos can seen from a particular portion of the Earth and it can be seen only at the time of coincidence of the particular place of the earth, with the planetary positions in the cosmos at a straight line. All the eclipses forming in the from time to time, normally at span of six months in the cosmos either in full or in partial according to the position of the earth, the shade planet, the Sun or Moon.

Lunar Eclipse occoured on 09.11.2003 Saturday

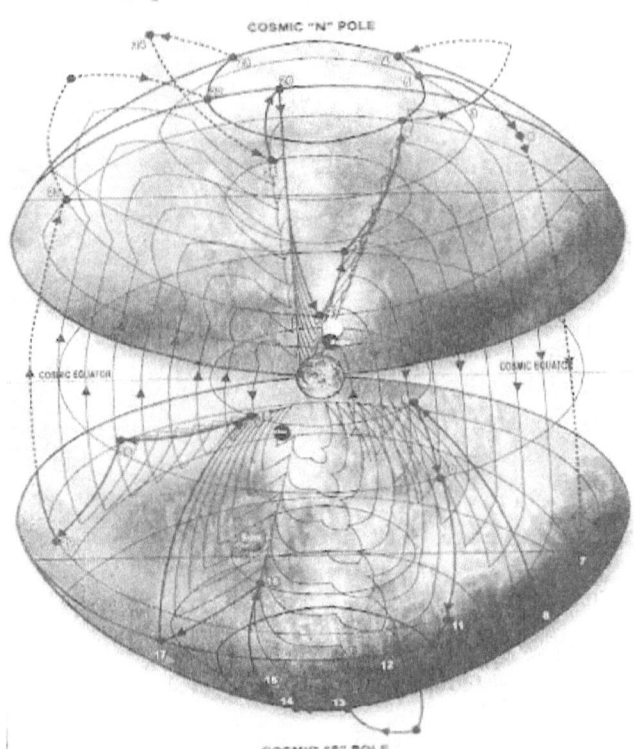

The Indian-cosmograph ABOVE provides the formation of Eclipse.

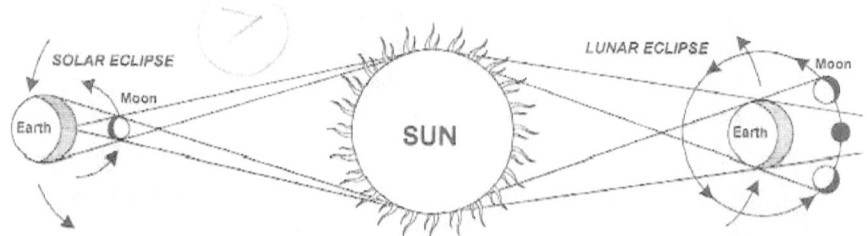

Eclipses as per Solar System (Western)

HELIOS – THE SUN CENTRED COSMOGRAPHY

Philosopher Kant, in 1755 A. D. expounded the theory of solar system, which systematized in 1796 by Lap lace on mathematical grounds. This system asserted that the Sun is stationery. Moreover, all planets revolve in orbits in the same plane, other movements being east to west. Keplar derived the laws regarding the motions of the planets. The phenomenon of eclipses and New moon also explained. May be these were the foundations to enunciate the coincidence of all orbits of the planets with equator in the same plane.

Ptolemy. An Egyptian Sailor and Astronomer were also in the opinion that the Earth is at the Center of the Cosmos. But the then Astronomers like Galileo, Aristotle, Gophernicloles, Kant and Keplar derived that the Sun in the center and the other planets including the Earth with its spinning movement and the Moon as its sub-planet to the Earth are revolving around the sun in the elliptical orbits in the cosmos.

Further, they discovered the new planets Neptune, Uranus and Pluto and included them beyond the position of the Planet Saturn in the cosmos. *They invented the Helios's system of cosmos and order of the planets by using the Ancient Indians' geo-centric cosmography.* It understood that they gave no importance to the stellar positions in the cosmos as if they too were unaware of the particulars about the integrated form of the stellar groups in the Universe at the time of their invention.

A closer study of the causes, which made them to develop a system against the existing one, the reasons, may be the following:

- They were in the opinion that the geo-centrism arrived on mere religious faiths and had no scientific reasoning.

- They had no belief on the explanations given by the religionist on the formation of the eclipses were based upon mere faiths.

- They wanted to prove the phenomenon of eclipses on scientific facts.

- They developed the Helios's by making some alterations in the order of the Planets against the order of the planetary system found in the Geo-centric system, where, the geo-centrist too felt that the planets are revolving in regular circular orbits around the earth in the Cosmos. They too had no idea about the exact location of the 27 stellar in the earth-centered cosmos and the stellar energy causes the movements of the planets and keeps the Earth at the center of the cosmos with a constant spinning movement.

- They knew, when the planets are in regular circular orbit, and then it might not be possible to prove the duration of time and the formation of eclipses to the actual occurrence of eclipses once in six in the cosmos. As if, the sun, moon and the earth are happen to be in straight-line at a frequent interval of 15 days of time in the Cosmos. To avoid the same they opted for the elliptical orbits around the Sun for the movements of the planets around the Sun.

They invented the Cosmograph, in which the Sun placed at the centre of the Cosmos. Further, they added the elliptical orbits around the Sun to of the planet founded in the geo-centre cosmography planets. They allotted the planetary positions around the sun by adopting the time taken by the each planet as detailed in the ancient Indians cosmography. If, they strictly adhere the periods of the planets, they would have placed the moons orbit near the Sun at it has only 27 days. As it seems impossible and they neglected the moons period and placed the Mercury's orbit next to the Sun which has 88 days, followed by Venus, the Earth in the place of Sun, Mars, Jupiter, Saturn, Uranus, Neptune and Pluto. As they found no place for the moon according to the sidereal periods, they assumed and allotted the moon as a sub-planet to the Earth and considered it motion around the Earth.

They rejected all the eight kinds of motions suffered by the planets in the cosmos, as they were derived mere superstitious illustrations. They proved the retrograde motions of the planets explained in the Ancient Indians Cosmography by explaining the movement of a train against the stationed train of plot form, a passenger in the stationery train has the backward movement of feeling when comparing the forward movement of the moving train. Further, they proved that the formation of eclipses in the cosmos once in six month, even the position of the sun, moon and the earth happens to be in a straight line once in 15 days. This is found due to their movements around the Sun in the elliptical orbits, where the planets may positioned at different locations in the cosmos when they transits through the Major and Minor axis portions of the elliptical orbits.

They explained that all the planets including the earth with its spinning movement and the moon as its sub-planet are moving in the same plane of elliptical orbits with the energy received from the Sun and proved the formation of eclipses beyond any doubt. These made the then astronomers to accept the Helios is which seems to be within the scientific facts and rejected the geo-centric cosmography. It is quite unfortunate that there is no refernce that the earth its sub-planet which has different types of motions. The earth has a spinning movement and a movement along the elliptical pathway around the sun and the Moon has the movement around the Earth and along the pathway along the elliptical pathway around the Sun.

It is found that a self rotating body which has a spinning movement along the elliptical way Cannot have a constant self-rotation when it transit through the major and the minor axis portion, where the varied forces can seen. If it moves, some disturbance must felt over the earth due to the varied spinning movement of the Earth, which causes the Gravitational force of the earth in the Cosmos. The Moon, which rounds the Earth, may also has the varied or disturbed movement due to force received from the Sun as there is no deviation in the gravitational force can seen, From this, it is realized that the twin planets have no chance to move around the Sun in the elliptical orbit. Likewise, no other planets can move around the Sun in the elliptical orbits in their positions.

Further, in the Helios's, it explained that the formations of Solar and Lunar eclipses are formed in the cosmos the position of the Moon and earth with reference to the Sun. The Node, which is the position of the moon or the shade of the earth or the position of the earth, which causes the eclipses and no reference, to the color variations seen in the eclipsed planets. It realized that the lunar eclipse formed due to obscure of the Earth's node or its shade. The Solar eclipse formed out of the Moon's position between the sun and the earth, and the color of the eclipsed planet always be either Black or Red in color which represents the earth or moon as the obscure matter, and have no chance to have the color variations as Red and Black in the Solar and Lunar eclipses.

A CRITICAL STUDY ON THE RELIABILITY OF THE BASIC PRINCIPLES OF THE HELIOS AND THE GEO-CENTRIC COSMIC THEORIES

Analysis on the reliability about the basic principles of the Present Helio's and the Ancient Indians' Oneness Universe and the Geo-centric cosmic theories based within the natural laws of Physics. I wish to submit that I am not favour of any particular theory of cosmology but my original intention is to emphasis which is very much reliable according to the science or real knowledge.

HELIOS' SYSTEM OF COSMOLOGICAL THEORIES

Since for the last two centuries that everybody in the world may know the basic principles and the assumptions were made on the basis of the scientific analysis and accepted as a real and set aside the Geo-centric cosmic theories as they were derived upon the religious and superstitious imaginations and not on the Scientific facts.

MOVEMENTS OF THE PLANETS IN THE HELIOS' SYSTEM

Now, It is noticed that the movements of the planets in an elliptical orbit around the Sun in their positions is also an assumption. But they have used all the sidereal periods of the planets movements along and parallel to the fixed and the integrated form of the 27 stellar groups which forms the egg-shaped cosmos. It is noticed that not only the westerners but also the Indians who practiced the stellar based movements of the planets during 476 C. E had also interpreted the movements of the planets around the Sun because they too had forgotten the fixed and the integrated form of the stellar groups in the cosmos and thereby had no idea about the movements of the planets based upon the stellar groups in the cosmos revealed by their Ancestors.

DO THE PLANETS CAN MOVE IN AN ELLIPTICAL ORBIT?

In the Helios' system it is assumed that an elliptical force line orbit has been generated from the Sun which makes the other planets to have their movements in their positions around the Sun.

Now let us consider the movement of the Earth and the Moon as its sub-planet around the Sun.

MOVEMENT OF THE EARTH AND THE MOON AROUND THE SUN:

When comparing the movements of the other planets around the Sun the movements of the Earth and the Moon is found to be a little complicated. Initially, The Earth has to move along the elliptical orbit generated by the Sun and it has to self rotates itself and the Moon has to rotate around the earth and along the elliptical pathway around the Sun.

FORMATION ELLIPTICAL PATHWAY FORCE-LINE

I hope that an elliptical force-line can be formed only if it has major-axis and minor-axis in which the force on the respective axis's must varied in their quantitative. If it is so, let us consider the movement of the Earth which has its uniform self-rotation throughout. This self-rotated Earth has a movement along the elliptical path-way around the Sun and it is taken as 365. 25 days to complete on round around the Sun. The Earth has to cross the major and minor axis portions of the elliptical path-way around the sun 4 times in a year and at an interval of 3 months. Hence, when the Earth passes through the Major-axis portions the self-rotating speed of the earth may have a set-back from its normal speed of self-rotation which may cause the difference in the gravitational force of the earth may be felt on the surface of the Earth at a frequent intervals of three months. But the reality shows that no such difference in the gravitational force of the earth has been felt. From this study we can asses that there is no chance to move parallel to an elliptical orbit around the Sun.

WHAT MADE TO ASSUME THAT THE PLANETS ARE MOVING AROUND THE SUN?

During 476 C. E an India Astronomer had stated that the planets are revolving around the Sun and the formation of Lunar eclipse is due to the interference of the Earths' Shade. It is not sure whether he paid attention to note the color variations in the Eclipsed planet at the time of Eclipse and further it is noticed during his time the fixed and the integrated form of the 27 Nakshtras groups which forms the Egg-shaped cosmos might not be available.

During 1755 C. E. when the Westerners made a study about the cosmic principles, in the absence of the fixed and the integrated form of the Nakshtras groups which makes the cosmos, they appreciated the Hindu Astronomical data's such as the sidereal periods of the planets i. e., the time taken by a planet to complete one round along and parallel to the fixed and the integrated form of the Nakshtras' groups and the occurrence of eclipses at frequent intervals. But they neglected and did not accept the Hindus 8 kinds of motions of the planets as they were revealed only by superstitious. Now, anybody can visualize the option for an elliptical orbit is nothing but to prove the formation of the Eclipses. If a uniform circular orbit of rotation Earth and Moon around the Sun is to be considered then there will be no possibility of forming the nodes at different positions as in the elliptical orbits and there may be no chance of formations of eclipses at an intervals of 6 months and at the same time it has the possibility of formation of eclipses at a frequent intervals of 15 days when the Sun, Moon and the Earth happens to be at a straight-line in the cosmos.

CONCLUSION

Ebenezer burgess, when he was in India during 1858, he translated the Indian Astronomical treatise and noted," Much yet remains to be done, before the History and use of the system of asterisms, as a part of the ancient Hindu Astronomy and Astrology shall be fully understood. There is in existence an abundant literature, ancient and modern, upon the subject, which will doubtless at some times provoke laborious investigation, and repay it with interesting results. We have already allotted to the Nakshatras more space than to some may seen advisable; our excuse must be the interest of the history of the system; as part of the ancient history of the rise and spread of Astronomical Science, the importance attaching to the researches of M. Biot, the inadequate attention hither to paid them, and the recent renewal of their discussion in the journals Des savants, and finally and especially, the fact that in and with the asterisms is bound up the whole history of Hindu Astronomy, prior to the transformation under the overpowering influence of the Western Science."

The developments and the inventions in the Mechanical, Agricultural, Industrial and other fields in the Western countries, made them to have fresh enquiries about the Astronomical Theories. During that time, most of the religions in the world whether knowingly or unknowingly followed the Geo-centric Cosmography. The then all-religious heads could not provide the proper reasoning about the geo-centric cosmographic theories, where the eclipses are forming in cosmos by the examples such as, swallowing of the planets by the Snakes and the biting of the Dogs. These made the then Astronomers with the curiosity to find the reasoning of the formation of Eclipses in the cosmos based upon the physical science.

The curiosity made them, to avoid the basic principles in the Geo-centrism. They are (i) The order of the planetary positions, (ii) The eight kinds of movements of the planets with reference to fixed and the integrated form of the Nakshatras in the Cosmos, (iii) The continuously generated energy from the Nakshatras causes the movements of the Planets and to upkeep the fixed position of the earth at the center and with a constant spinning movement. (iv) The shade planets formed over the surface of the earth due to the penetrated rays of the nakshatras through the planets Saturn, Mars, and Jupiter, which causes the occurrence of the eclipses in the Cosmos.

By considering the sidereal periods of the planets to complete one round along and parallel to the fixed Nakshatras of the planets in their positions, they invented the Helios's system of cosmography, which had the criticisms by the then scholar Ebenezer Burgess. It revealed, it is quite unfortunate to the then astronomers as they have no opportunity to have the details of the fixed and the integrated form of the Nakshatras which forms the cosmos around the Earth. Which lead them to think otherwise even without briefing about the critical movements of the twin planets Earth and the Moon, which have their movement along the elliptical orbit around the Sun

In fact, it is not the aim to find the flaws in the system. But to high-light the neglected or a dead system found to be worth in the laws of science, by evolving the position of the fixed and the integrated form of the Nakshtra groups, which forms the Cosmos by using, The Surya siddhantha and the Indian Almanac. In modern times, it is understood that most of the Scholars are behind mere in faiths by considering the mere distance of the locations of the Nakshatras groups from the earth and not ready to accept that the Nakshatras energy causes the movements of the planets in the cosmos on scientific grounds.

It is found in India that the Science based Ancient Indians Geo-centric Cosmography transformed into religious base and observed by the Indians in their all lifestyles! In India, an auspicious or inauspicious events, they start their events, by narrating the cosmic details like, the name of the current Kalpa, Manvantara, Yuga, Geographical location, the name of the year among the 60 cycle of years, the name of the Month, the name of the Paksha, the name of the week day, the Nakshtra group, through which the moon transits, the names of their ancestors and their relation-ship to the particular individual, his name and the Nakshatra group when he took birth. But, some of the Indians too have no belief over the practice of the Ancient Indians' geocentric cosmic theories as if they could not compared them with the existing Helios's system of Astronomy in which all the planetary positions had changed.

In India, all the Astronomical events are practiced in the form of rituals in the daily life. Special events, such as the formations of eclipses in the cosmos, the Full and New moon periods, the transit period of the Sun from one constellation to the other, and the transit period of the planets from one constellation to the other constellation. In all above events, in general, the cosmic energy which acts around the earth causes the desired and undesired actions over the men and matter on the earth, to set-right them by the Dos and Don'ts. To keep the Do's and Don'ts by the illiterates they provided the examples such as the snake sallow the planets at the time of eclipses and the certain thing should not carried out.

In the last many centuries, the followers of Earth centred cosmic they could not explain to the then enquirers, in scientific ways about the occurrence of the eclipses, the retrograde motions of the planets and the opposite movement of the shade planets Rahu and Kethu in the cosmos, as they were not able to produce graphical representation about the Fixed and the integrated form of the Nakshatras groups around the Earth in the cosmos.

Now it understood with the evaluation of the Fixed and the integrated form of the 27 nakshatras groups with reference to the Earth's equator in Ascending and Descending order which forms the Oneness Universe. The energies released by the nakshatras group within the Cosmic-shell, causes the movements of the planets in the Cosmos and stationery position of the Earth with a constant spinning movement in the Cosmos. I request the Scholars and the Astronomers to review the same, if it done without any bias, which will do the betterment of the Men and matter on the Earth and assured success in launching satellites to the desired Planets from the Earth.

BIBLIOGRAPHY

1. Surya Siddhhantha –A textbook of Hindu Astronomy-Burgess Ebenezer

2. Critical study of ancient Hindu Astronomy-Somayaj D. A

3. Eclipse cult in the Vedas, Bible, and Koran-A supplement to drpsa-Sahastry.

4. Rudimentary Astronomy-Main (Robert)

5. Study of Solar System-Chamers George F)

6. Lessons in Astronomy including Uranography-Young (Charles A)

7. Astronomy has the earth movements Detailed Study-(Subramanya Sastri V)

8. Elements of Descriptive Astronomy-Tancock

9. Indian Ephemeris and Vakya panchangas.

10. Tantra Vidya based on Archaic Astronomy and Tantric Yoga Translation from original German-Bedekar (Hize Oscar Marcel)

ANNEXURE

ANNEXURE-I

Positions of the plantes at Thanjavur. The following details showing the Planetary positions which are shown in the figure as on 20. 08. 1953

1. **Sun at Maham 2nd Padam 123°**
2. **Moon at Moolam 4th Padam 250°**
3. **Mars at Ayilyam 2nd Padam 110°**
4. **Mercury at Ayilyam 4th Padam 116°40'**
5. **Jupiter at Mirgasheer 2nd Padam 56°40'**
6. **Venus at Punarpoosam 2nd Padam 83°20'**
7. **Saturn at Chittirai 2nd Padam 176°40'**
8. **Rahu at Uthiradam 4th Padam 276°40'**
9. **Kethu at Poosam 2nd Padam 96°40'**

Lunar Eclipse occurred on 9. 11. 2003 Saturday. Eclipse is due to the interception of Rahu in between the line of vision from the Earth to Moon.

Sun Visakam 1 Padam 201°

Moon Bharani 3 Padam 13°20'– 26°40'

Rahu Bharani at 3 Padam 20°

Kethu Visagam 2 Padam 200°

ANNEXURE–2

Detailed Discussion forum and webpage of the author:

https://swamycosmology. wordpress. com/

ANNEXURE-3

Author's studious effort for including the Indian Panchang in education system has made him to appeal Public interest litigation in front of the judiciary in Honourable high court of Madras and has received a feedback in favour of his views.

IN THE HIGH COURT OF JUDICATURE AT MADRAS

DATED : 23.02.2018

CORAM

THE HON'BLE MS. INDIRA BANERJEE, CHIEF JUSTICE

AND

THE HON'BLE MR.JUSTICE ABDUL QUDDHOSE

W.P.No.3879 of 2018

R.Swaminathan ... Petitioner

Vs.

1 The Chairman cum Managing Director
   Satellites Launch Services and
   Tracking Facilities, Antariksh Bhavan
   Bell Road Bangalore-560 231.

2 The Chairman
   National Council for Research and Training
   Sri Arabindo Marg, New Delhi-110 061.

3 The Secretary
   Central Board of Secondary Education
   New Delhi-110 061.

4 The Government of India
   Rep. by the Secretary
   Ministry of Education
   New Delhi-110 061.

5 The Chairman
   Tamil Nadu Text Book Society
   College Road
   Chennai-600 006.

6 The State of Tamil Nadu
   Rep. by the Secretary
   Department of Higher Education
   St. George Fort Chennai-600 009.

7 The State of Tamil Nadu
   Rep. by the Secretary
   Department of Education
   St. George Fort,
   Chennai-600 009.   ... Respondents

CI0077924

63

PRAYER: Petition under Article 226 of the Constitution of India for issuance of a writ of mandamus directing the respondents to implement the study in unit at the research levels and the Indian Panchang at Primary and Secondary levels in our educational Institutions pursuant to the representation of the petitioner dated 11.09.2017.

For Petitioner : Mr.P.Raghunathan
for M/s.T.S.Gopalan and Co.

For Respondents : Mr.C.Munusamy
Spl. Government Pleader
for respondents 5 to 7

ORDER
Order of the Court was made by Ms.Indira Banerjee, Chief Justice]

After hearing learned counsel appearing on behalf of the petitioner and on perusal of the writ petition and the documents appended to the typed set of documents filed on behalf of the petitioner in support of the writ petition, we dispose of the writ petition by directing the respondent Nos.2, 3, 4, 6 and 7 to consider introducing Ancient Astronomical and Scientific Treatises as subjects of study in educational institutions in accordance with law. No costs.

Sd/
Assistant Registrar

/True copy/

Sub Assistant Registrar

To:

1   The Chairman
National Council for Research and Training
Sri Arabindo Marg, New Delhi-110 061.

2   The Secretary
Central Board of Secondary Education
New Delhi-110 061.

3   The Secretary
Government of India
Ministry of Education

c 0078981

ANNEXURE-4

60 Years in Indian calendar sysytem-samvastaram

01. Prabhava 1396–1397 1162–1163 2018 1987–1988
02. Vibhava 1397–1398 1163–1164 2019 1988–1989
03. Sukla 1398–1399 1164–1165 2020 1989–1990
04. Pramodhoodha 1399–1400 1165–1166 2021 1990–1991
05. Prachorpaththi 1400–1401 1166–1167 2022 1991–1992
06. Aangirasa 1401–1402 1167–1168 2023 1992–1993
07. Srimukha 1402–1403 1168–1169 2024 1993–1994
08. Bhava 1403–1404 1169–1170 2025 1994–1995
09. Yuva 1404–1405 1170–1171 2026 1995–1996
10. Thaadhu 1405–1406 1171–1172 2027 1996–1997
11. Eesvara 1406–1407 1172–1173 2028 1997–1998
12. Vehudhanya 1407–1408 1173–1174 2029 1998–1999
13. Pramathi 1408–1409 1174–1175 2030 1999–2000
14. Vikrama 1409–1410 1175–1176 2031 2000–2001
15. Vishu 1410–1411 1176–1177 2032 2001–2002
16. Chitrabaanu 1411–1412 1177–1178 2033 2002–2003
17. Subaanu 1412–1413 1178–1179 2034 2003–2004
18. Thaarana 1413–1414 1179–1180 2035 2004–2005
19. Paarthiba 1414–1415 1180–1181 2036 2005–2006
20. Viya 1415–1416 1181–1182 2037 2006–2007
21. Sarvajith 1416–1417 1182–1183 2038 2007–2008
22. Sarvadhari 1417–1418 1183–1184 2039 2008–2009
23. Virodhi 1418–1419 1184–1185 2040 2009–2010
24. Vikruthi 1419–1420 1185–1186 2041 2010–2011

25. Kara 1420–1421 1186–1187 2042 2011–2012

26. Nandhana 1421–1422 1187–1188 2043 2012–2013

27. Vijaya 1422–1423 1188–1189 2044 2013–2014

28. Jaya 1423–1424 1189–1190 2045 2014–2015

29. Manmatha 1424–1425 1190–1191 2046 2015–2016

30. Dhunmuki 1425–1426 1191–1192 2047 2016–2017

31. Hevilambi 1426–1427 1192–1193 2048 2017–2018

32. Vilambi 1427–1428 1193–1194 2049 2018–2019

33. Vikari 1428–1429 1194–1195 2050 2019–2020

34. Sarvari 1429–1430 1195–1196 2051 2020–2021

35. Plava 1430–1431 1196–1197 2052 2021–2022

36. Subakrith 1431–1432 1197–1198 2053 2022–2023

37. Sobakrith 1431–1433 1198–1199 2054 2023–2024

38. Krodhi 1433–1434 1199–1200 2055 2024–2025

39. Visuvaasuva 1434–1435 1200–1201 2056 2025–2026

40. Parabhaava 1435–1436 1201–1202 2057 2026–2027

41. Plavanga 1436–1437 1202–1203 2058 2027–2028

42. Keelaka 1437–1438 1203–1204 2059 2028–2029

43. Saumya 1438–1439 1204–1205 2060 2029–2030

44. Sadharana 1439–1440 1205–1206 2061 2030–2031

45. Virodhikrithu 1440–1441 1206–1207 2062 2031–2032

46. Paridhaabi 1441–1442 1207–1208 2063 2032–2033

47. Pramaadhisa 1442–1443 1208–1209 2064 2033–2034

48. Aanandha 1443–1444 1209–1210 2065 2034–2035

49. Rakshasa 1444–1445 1210–1211 2066 2035–2036

50. Nala 1445–1446 1211–1212 2067 2036–2037

51. Pingala 1446–1447 1212–1213 2068 2037–2038

52. Kalayukthi 1447–1448 1213–1214 2069 2038–2039

53. Siddharthi 1448–1449 1214–1215 2070 2039–2040

54. Raudhri 1449–1450 1215–1216 2071 2040–2041

55. Thunmathi 1450–1451 1216–1217 2072 2041–2042

56. Dhundubhi 1451–1452 1217–1218 2073 2042–2043

57. Rudhrodhgaari 1452–1453 1218–1219 2074 2043–2044
58. Raktakshi 1453–1454 1219–1220 2075 2044–2045
59. Krodhana 1454–1455 1220–1221 2076 2045–2046
60. Akshaya 1455–1456 1221–1222 2077 2046–2047

Notes

****

Notes

****